Stargazing

What to Look for in the Night Sky

Tom Van Holt

Illustrations by Greg Hardin

STACKPOLE
BOOKS

For Max and Brooks

Published by
STACKPOLE BOOKS
5067 Ritter Road
Mechanicsburg, PA 17055
www.stackpole books.com

Printed in the United States

10 9 8 7 6 5 4 3 2 1

First edition
Cover design by Wendy Reynolds.
Cover photograph of Comet Hyakutake by Robert Sandy.

Library of Congress Cataloging-in-Publication Data

Holt, Tom Van.
Stargazing / Tom Van Holt.
p. cm.
ISBN 0-8117-2934-6
1. Astronomy—Observers' manuals I. Title.
QB64.H65 1999
520—dc21 98-30713
CIP

CONTENTS

ACKNOWLEDGMENTS

Thanks to Stephen Shawl, professor of astronomy at the University of Kansas, whose dedication to the mystery in the sky is matched only by his devotion to his students, among whom the author is proud to count himself.

To Greg Hardin, who gave far more than anyone had a right to ask—outstanding!

To Jackie Wade, of the Astronomical Society of Kansas City, who would share the whole universe.

To Dave Lindsay, National Radio Astronomy Observatory; Patrick McCarthy, U.S. Naval Observatory; Rick Clements and Ann Hyde, Spencer Research Library; Bruce Bradley, Linda Hall Library; Norris Heatherington, University of California at Berkeley; and David Tracewell, illustrator, my cousin.

And to the Writer's Group, for encouragement, camaraderie, and great dinners.

INTRODUCTION

In this book you will find humor, outdoor survival techniques, mind-altering concepts, history, great discoveries, old-wives' tales, believe-it-or-not stories, and the forces of nature reduced to something not a whole lot more complex than what makes your washing machine spin. It is for people intrigued by the entire mystery of the sky above, of which science is only one part.

You will not find lots of numbers followed by lots of zeroes, involved discussions of things you'll never see that will make no difference in your life, and tedious instructions that you'll forget immediately. With few exceptions, everything in this book can be seen by a person with the naked eye just a few miles from city lights. It's point of view is that of people living in the Northern Hemisphere between latitudes 30 and 50 degrees, essentially between upper Mexico and lower Canada. The purpose of this book is to decisively confirm the beauty, mystery, and power every person feels when peering into the night sky.

RHYTHMS AND PATTERNS

There's a secret to understanding the sky at night. It won't be found within the inner mechanisms of a telescope nor within the hallowed hallways of higher learning. This secret is patience, the quiet magic needed to acquire any skill.

Learning about the sky at night is much more intimidating than learning to speak Spanish or build a chair of wood. Those pursuits have simple, tangible building blocks that bring one slowly toward mastery. With stargazing it's different. Your classroom has been moved outdoors; it has become the universe. It is a dynamic entity that changes through the year and by the hour and always lies far beyond reach.

The universe is a big place. There's no end to it and what can be learned about it. Though every man, woman, and child on Earth share the stars, only a fraction of a percent know anything about them. Yet nothing has had more impact on our lives than the simple act of people looking at the sky. "A person deprived of the broad outlines of astronomical knowledge is as culturally handicapped as one never exposed to history, literature, music or art."* Far from being the sacred preserve of academicians or cultural aesthetes, stargazing has been the premier skill of some of history's most daring individuals: discoverers, soldiers, magicians,

* Dave Finley, public relations director, The Very Large Array, National Radio Astronomy Observatory in Socorro, New Mexico.

The stars have inspired and guided the world's greatest adventurers, from pirates to pioneers.

and rebels. They learned about the stars in the same way you will—by looking up, and by doing it often.

Stargazing is not as immediately dramatic as sports like kayaking or hunting. It will not inspire as much awe at parties as popping the wine smartly or jabbering about your new car. To enjoy the stars properly, you may be forced to take a twenty-minute drive to a quiet country location, walk a short distance, and lie on your back for half an hour. It takes time and patience. You will learn about the one part of history that will never become obsolete and be a quiet witness to events of great drama and moment.

In one evening, it is possible to both observe and understand the majestic clockwork of the sky. By no means will this be a dry evening of tedious self-instruction. Several shooting stars and satellites will streak by, the Milky Way will emerge from the blackness, and planets will appear among the creatures of the zodiac. Guaranteed. Like good music, it's mildly appealing at first, and then it grows on you. And though the music of the spheres always varies, the underlying rhythms and patterns never change.

The motions that order the stars and planets are easy enough to recognize, once you understand them. This chapter will explain them, along with all the other things that make a regular appearance in the night sky. But the universe is a big place, and these rhythms and patterns are only a beginning.

Stargazing is the foundation of all science and plays a huge role in history. It takes time to become familiar with the whole universe, but the knowledge you acquire will be valuable for your entire lifetime. The planets will not alter their course; the stars will not blink out.

To find north, to distinguish one pinpoint of light from another, or to be able to adapt a flat representation on paper to spherical reality all take practice. The pell-mell motions of celestial objects are confusing to a person casually admiring the starry night, even to an experienced outdoorsman. Watching the stars is a lot like watching a football game: nothing makes sense until you know the rules. On one level, the rules of the universe are a lot simpler than those of football. It is reassuring to know that the stars steer so sharply on course that primitive men and modern scientists have used them as the organizing principles for civilization's greatest accomplishments.

SEEING IN THE DARK

Let's forget about the mysteries of football for now and wander around in the dark instead. Though it's helpful to escape the lights of the city, a place deep in the country is not essential. In fact, the country sky can be so rich with stars that it makes it difficult to find a particular one. It's not necessary to see thousands of stars, just a few hundred. Getting a few miles past the lighted streets of town should do. Find a place with an open horizon in as many directions as possible, the higher the better. Leave as the sun is setting, when you will be treated to the colorful, time-between-worlds of twilight. Then, too, the motion of the distant stars will be deeply impressed against Earth's fading silhouettes.

If conditions allow you to lie or sit down, by all means do so. Plan on staying a while and make yourself comfortable. Take along a blanket or a sleeping pad, even a pillow. Take this book. Take a flashlight (it's less blinding if the lens is covered with red tape or Magic Marker) and a map of the area, particularly a map that indicates terrain features, if possible. This will help you start learning to associate places with direction and celestial objects. Take a compass. A $2 crackerjack will do, and don't worry, it's only necessary to know that the red half of the arrow points north. If you can already find north or the Big Dipper, you won't need one.

It's a good idea to find a favorite place and stick with it for a while. This way you're able to focus on the changes taking place in the sky, not on Earth. Then it's obvious that each star always rises and sets in exactly the same place, and that the planets and the moon always rise and set within the same narrow band. Knowing a hill or pond over which a star can always be seen cresting makes the universe a part of the world you know. Get an idea where

To observe the night sky, get comfortable, take your time, and bring a friend.

familiar landmarks—such as a pond, a grove of trees, or distant city lights—are and how they lie in relation to each other.

SEEING STARS

Look Where the Sun has Set

The east-west arc that the sun travels through the sky each day migrates with the seasons: south in the winter, north in the summer, just like the birds. After the summer solstice (around June 21), the arc moves southward. If you note the place where the sun rises or sets each day, you will see this point slowly creep along the objects on the horizon as the days go by. This is why the shadows in winter become so long, the days so short, and sunset seems forever imminent. It's worth noting, because this change in our angle to the sun is why we have the four seasons. It's so easy to observe that it can be seen while commuting or simply stepping outside and observing where the sun is peaking over the roof. In ten days, the motion will become quite apparent.

After the winter solstice (around December 21), the pathway migrates northward to a position higher in the sky. The sun's rays are striking us more directly and making the temperature increasingly warmer. The sun is the only star you can see move independently in the sky, and this apparent motion doesn't affect the motion or position of anything else.

Look Opposite the Sun, to the East

This is where every star makes its nightly debut. You may find it easier to use your peripheral vision to pick out fainter stars or bits of color, such as the red of the planet Mars or the yellow of the star Sirius. Why? The optic nerve linking the eyes to the brain connects directly behind the pupil. In doing so, it displaces the light-sensitive cells that line the inside of the eyeball. By looking at objects not quite so directly, you take full advantage of your eyes' strength.

A useful way to locate objects in the night sky is to use your hand as a ruler. This method has been used for at least three thousand years—it likely will work for you, too. Hold your fist at arm's length, and sight over the back of your hand. This spans an arc of roughly 10 degrees (one finger is about 2 degrees). With a little practice, you will know exactly the width of your hand. This method is used a great deal in this book, and you will find yourself naturally using it to point things out to others.

You can use the fist method to find the one part of the sky that never appears to move; it is around this point that the entire heavens seem to turn. It hovers above the axis that the Earth spins on, and you can find it with your compass. Face directly north, then sight four fist widths above the horizon (roughly 40 degrees). The star sitting on top of your finger is Polaris, the North Star (page 8). This is true north. The best compass in the world gives a less honest north than this star. Though not very bright, it has guided people for hundreds of years.

The North Star barely moves and never sets, just as the axle of a wheel can be seen to spin but not change place. There are many more stars that circle tightly around it, and they never set either. These stars, along with Polaris, are known as the circumpolar stars. You will learn to cherish them and the constellations they form as your loyal guides to the sky. They are easy to see and always visible, pointing the way to other stars.

When the stars appear to set, it's because the Earth is spinning us away from them, eventually taking us back into the blinding rays of the sun. The stars *are* moving through outer space, and at incredible speeds. But they are so vastly distant that we can't see this movement except over hundreds and thousands of years. Be grateful for this optical shortcoming, because if the stars could be seen speeding in their thousand different directions, we'd perpetually feel seasick. The only reason we can see the sun move in relation to the other stars is because it's 300,000 times closer than the next closest one.

The North Star is "four fists" above the horizon.

The sun is a star, and it's unusual only because it happens to be *our* star. Everything that can be found in a planet is in a star, and more—more different elements, and millions of times more of them. This makes stars so massive that they put tremendous pressure on their insides. This is similar to the pressure that occurs if you dive into deep water—the column of water above you gets taller and heavier, and it can make your head hurt. But with a star, like our sun, the pressure is so great that it creates intense heat. When you inflate your car tire, the increase in air pressure heats up the tire's inside. But the air is cool when it hisses out—it cools as it escapes and the pressure is eased.

The stars are not only at tremendous distances from us, they are at a tremendous range of distances from us, all with vastly different sizes and brightnesses. Constellations that appear to be composed of several similarly bright stars actually have some that are near but dim and others that are far but bright. Seen from another vantage point in the galaxy, those stars might not come together as constellations at all. It is only by coincidence that we have the number of striking sky-pictures that we do, for chance does not ordinarily allow such beauty. This is but one of the many coincidences that inspired the ancients to believe there was magic in the sky.

As the Earth revolves around the sun, we are able to peek into a new little slice of the universe each night—and another little slice passes from view. We see more and more as we journey along our circular path around the sun, until ultimately we have made a full tour of our arm of the Milky Way galaxy each year, at a speed of 70,000 miles per hour! Our stargazing would be much simpler if there were no sun, for it's always blinding us to half the view. If we could just shut it off for a day, we could see all the galaxy in just one twenty-four-hour turn of the Earth (by which time, however, we would have frozen to death).

What this revolving around the sun means to the stargazer is that the stars rise four minutes earlier each night. But they never change where they rise, nor do they change in relation to each other. They will advance deeper and earlier into the night, so that a star that had just been rising at 9:00 P.M. is well clear of the horizon at the same time a week later. It then becomes necessary to stay up later and later to keep the same ones in view. Eventually they'll be absent for a few months from our night sky. This is why stars visible at 11:00 P.M. in March can be seen in the same position at 9:00 P.M. in April, 7:00 P.M. in May, and not at all by June. The stars are always there; it's just that we may not be turned away from the sun so that we can see them at a particular time.

This movement of the Earth—spinning on its axis while moving in a great circle around the sun—is entirely what accounts for the apparent motion of the stars. Earth's movement is similar to the octopus ride at a carnival. The octopus spins its arms in a circle, and the cars at the ends of the arms also spin on their own. The marquee lighting on the octopus's head blinds you to anything in that direction. Not only does it make you sick, but it means that you can't see your friends below until your car has both spun outward and the arm has brought them within sight.

THE SOLAR SYSTEM

Unlike the stars, the sun, moon, and planets are not fixed in place. We are so much closer to them that we can see them moving against the background of stars. It's like watching a kite bobbing and weaving across the sky while a distant jet can hardly be seen to move. Each planet maintains its own speed and course, which makes finding them a little tricky. The word *planet* comes from the Greek, meaning "to wander."

The individual pathways of the sun, moon, and planets are as well rutted as that of any star. Even better, all their paths lie within a narrow band of sky, and all of these bodies move in the same direction. The moon goes counterclockwise around the Earth (as seen from above the North Pole), and all the planets and Earth go counterclockwise around the sun. And because they lie in nearly the same plane as the sun, they follow nearly the same path

Earth's movement is similar to the octopus ride at the carnival.

through the sky. This is a valuable clue for planet spotting: where you see the sun in the day, so, too, will you find the planets and moon by night. This is why we frequently see spectacular alignments of the planets and moon (it's not nearly so coincidental as astrologers suggest).

Even though the planets are much harder to see than the sun and moon, they travel much more slowly across the sky. For example, once you've found Jupiter, you can easily track it for months afterward. It's possible to see five of the planets with the naked eye if you have good or corrected eyesight. These are the same planets that people have watched forever: Mercury, Venus, Mars, Jupiter, and Saturn. Soon you'll learn to see them, too.

They don't lie exactly in the same plane; if they did, every time a planet or the moon passed between the Earth and sun we'd have an eclipse. (This is why it's called the "plane of the ecliptic.") The band the sun, moon, and planets travel within is only about one and a half fists wide. They were created together out of a vast, swirling cauldron of gases, and they haven't abandoned that swirling just because they've condensed into heavier bodies. Nonetheless, it is a very flattened swirl.

Why haven't the planets abandoned their spawning grounds and hurtled into deep space? Why doesn't mighty Jupiter, so huge and distant from the sun and flying through space at 30,000 miles per hour, tear off on its own course? What holds the solar system together? You know the answer: gravity.

What is gravity? You know the answer as well as the most acclaimed physicist, for no one knows exactly what gravity is. We only know that it rigidly dictates the workings of the entire universe, from a spinning ballerina to a juggernauting cluster of galaxies. The larger the body, the greater the gravitational force within its domain, such that our galaxy holds the sun whirling in orbit, the Earth holds the moon captive at its side, and the moon draws the tides out of the oceans.

The ancients were ignorant of gravity as a force moving the heavens. Yet they were confident enough of that movement that

Though the naked eye has revealed only five planets to people, we have always suspected that there are far more.

they planted and navigated by the stars. So today are we ignorant of why gravity is a prime mover, but we know that its law is so unbreakable that we can send an unsteerable rocket to the moon; if our calculations are correct, gravity will unfailingly bring them to their appointed meeting place.

It's a lot to grasp, and it takes patience to visualize these motions and forces in the sky. But you are not alone. It's taken thousands of years for us to understand the mechanics of the solar system, and if our view of the sky has improved, it's only because we stand on the shoulders of men who stood on the shoulders of men standing in a great pyramid reaching down to the beginning of history. People didn't have to know why the sky moved; they only needed to learn the rhythms and patterns it unfailingly traced out. Few of us understand the gasoline engine, yet our very way of life depends on it.

As of this moment, you know more about the universe than everyone in the world before 1687. Surprise. The puzzle of sky motion was that difficult for people to figure out. And you learned it in less than a half hour's time. Long ago, people used those reliable rhythms and patterns—the only ones they really had—to order the most important parts of their lives. Now that you understand the workings of the great timepiece, you need only watch to see the gears, springs, and rockers move in machinery so huge that it takes time merely to see it from one end to the other. Be patient.

What, now, besides the stars, sun, moon, and planets, can the naked eye always see in the night sky? A ghostly trail the width of your outstretched hand, reaching from one horizon to the other: the Milky Way. It's not the pathway of souls to heaven or the road to Mount Olympus, as some believed, but the galaxy we inhabit. Our galaxy is a great spiral, which we are seeing on edge from our vantage point two-thirds of the way from the center. The band of haze we see is made up of faraway stars without number, appearing about one fist width wide. The Milky Way by itself is so huge that when we look at the sky, every star we see is no farther away than it's closest arm. The rest of our galaxy and the universe, with

Thousands of great minds have lifted us to our present understanding of the night sky.

the exception of the Andromeda galaxy (pages 99–100), is simply too far away to be visible. If you want to get a good look at the Milky Way, you'll need to get away from city lights to a darker location. It's well worth the trouble, for it's truly a spectacle. It can't be missed—it's by far the largest feature in the sky and never changes its location among the stars.

SUMMARY

- The sun always rises in the east and sets in the west, in an arc that heads south for the winter and returns north for the summer.
- The Earth spins beneath the North Star, causing everything in the sky to appear to turn from east to west in a circle around it. Each star always rises and sets in the same place and travels the same arc across the sky. The stars do not change in relation to each other. It is only the time they appear that changes.
- All the members of the solar system—the sun, the moon, and the planets — lie on nearly the same plane. For this reason they move through the same path in the sky, a path most obviously journeyed by the sun every day. Because they are so close to us, they can be seen to move independently of the stars and each other.
- The Milky Way galaxy is our home, and we are surrounded by the stars in it. We see a side view of this hazy band reaching from one horizon to another, about the width of an outstretched fist.

WANDERERS IN THE SKY

Often the most interesting objects in the night sky are those that are in motion or are irregular visitors. This very motion and irregularity can make them easy to find yet hard to identify. Knowing what you're looking at may require the best viewing conditions possible, a quiet place far away from lights and pollution.

This chapter describes how to find all the objects that can be seen moving in the night sky and how to tell them apart from each other. You may look into the sky and see as many as 8,000 stars, but all the other objects you see will never amount to a hundredth of that number—and they're moving!

How do you find a satellite, shooting star, or planet? Two things are certain if you see something move across the sky: 1) It's within our solar system; and 2) it's not a star (except for the sun). The stars are too distant and the universe too vast to detect their motion with the naked eye, even if they're zipping along at a million miles per hour.

METEORS

Of all the lights we see in the sky, the most dramatic and common are the shooting stars, or meteors. These are the flashing streaks of light we wish upon or cheerily regard as "another soul going to heaven" (though many are pointed downward). But these meteors are just a flash in the pan, consumed by the very blaze of their debut.

Each day, millions of trillions of meteoroids survive their blaze through our atmosphere and strike the Earth, now mete*orites*. Most of these are smaller than a grain of sand. On the other hand, one has left a gaping hole in our own United States—Meteor Crater, just east of Flagstaff, Arizona, over a mile across and a third as deep. It's so large that you can barely see the remains of a plane that crashed inside over thirty years ago.

As impressive as Meteor Crater is, it's tiny compared with others. There's speculation that Hudson Bay is an ancient impact crater, and evidence has recently been presented suggesting that this was how the Gulf of Mexico was formed. There are huge impact sites all over the Earth, but most are so old and large that they've become a part of the landscape and now can only be seen from space.

Meteor Crater, near Flagstaff, Arizona.

Scientists and collectors have found the Antarctic to be the best place to see and locate meteorites.

Though most meteorites are paperweight or dust-speck size, occasionally one bursts into a bedroom or plows through a car. The extraterrestrial metals that they bring with them were used for the tools and weapons of aboriginal peoples worldwide, especially in the drier climates, such as Alaska or the desert Southwest. When scientists want to find meteorites, they do it where the conditions are somewhat similar to the moon: in the Antarctic. Here they won't be washed away or buried, can be spotted easily, and are found in such abundance that entrepreneurs can collect enough to sell at a profit.

Meteors can actually be caused by any piece of space debris entering our atmosphere—man-made space junk, dust, a disintegrating comet, or bits of a planet's rings. But their most common source is from a place few of us have ever considered: a missing planet.

When Renaissance astronomers began applying their deductive logic to the universe, they noticed that the orbits of the planets were spaced according to a simple pattern they called Bode's Law.* But there was a gap between Mars and Jupiter that this spacing theory couldn't account for. Then astronomers slowly discovered the asteroid belt—a primary source of our meteors. This debris could have become a planet if not for the gravitational forces of titanic Jupiter which ripped it apart.

Some asteroids are huge like mighty Ceres, which is 400 miles across, but most are far smaller. These lightweights may be easily nudged out of the belt by a gravitational tug as slight as that of a passing star trillions of miles away. Eventually this debris may be drawn to Earth or any other object in the solar system. Like an Apollo reentry vehicle, they flare up on their 100,000-mile-per-hour descent through our atmosphere, making them brightly visible.

* Bode's Law states that the planets are spaced according to this formula: add 4 to the sequence of 0, 3, 6, 12 . . . , then divide each by 10. This tells how far each planet is from the sun in Earth-sun distances.

Seeing meteors requires but one thing: patience. On almost any clear night, you can expect to see at least one within twenty minutes, anywhere in the sky. With practice and a good location, you can see dozens in an hour.

Several times a year we are treated to meteor showers, which can be virtual rainfalls of streaking lights bursting from the same region in the sky. Some showers are random and unpredictable, occurring when the Earth encounters a swarm of meteoroids in its journey through space. Others occur on a very regular, predictable basis. These would better be called "comet showers," because this is their true source, not the asteroid belt. A comet may begin to deteriorate, and instead of one distinct distant passerby, we are treated to a hail of debris pulled out of the disintegrating comet's decaying orbit. The best comet showers are the Geminids, visible every year around December 13, and the Perseids, visible around August 11. On these nights, you are guaranteed to see dozens of meteors, some of which resemble a spacecraft coming down in flames.

COMETS

A comet is not a shooting star, nor is it on fire. It's a big, BIG snowball that's visible only because the sun is shining on it. Though the vaporous head of a comet may be over 500,000 miles across and its tail may extend 100 times farther, it has a tiny body, maybe 5 miles across, which is nothing more than frozen gases, dust, and metals. A comet has almost no mass whatsoever but puts on a great show as it approaches the sun, for it heats up and essentially steams (like dry ice) into a huge, gaseous ball. The vast tail always points away from the sun, the force of the solar wind and light rays blowing the steaming material backward.

Comets may be millions of miles away, barely visible as more than a wisp of fog. Or they may be visible in the day, as with the famous Halley's comet of 1910, when Earth was almost swept up in its tail. Comet Hale-Bopp of 1997 and '98 was just the right distance to be magnificent yet not inspire worldwide panic, as many comets have.

The size, speed, duration, and orbit of each are different, though an individual comet assumes predictable orbital characteristics with each return trip. If one is visible, there will be plenty of media attention. Comets may linger for months, as did three recent ones, Hale-Bopp, Hyakutake, and Halley. Comets are best observed when they are close to the sun; not only is it their only source of illumination, but they're in orbit around it. So it's best to look for them just after sunset and just before dawn, close to the hidden sun.

Many astronomers believe that there is a huge sphere of comets, called the Oort Cloud, surrounding the solar system and filled with billions of racing comets that are occasionally tugged out of the cloud as a star or planet pulls on it. Until they lose speed or gas out altogether, comets will endlessly repeat their immense journey around the solar system, some circling in a few years, others requiring thousands.

ARTIFICIAL SATELLITES

The Global Positioning System (GPS), increasingly popular with outdoor lovers, is made possible by twenty-seven military satellites. These are among the two hundred visible satellites of the six thousand flying overhead. They all lurk above Earth anywhere from 3,000 to 20,000 miles distant. The GPS is but the latest of their many applications: environmental watchdogging, mapmaking, stargazing, telecommunications, and so on.

Satellites have only been around since 1959, when the Russians launched *Sputnik* and started the race to conquer space. Each one has to be sent up in a rocket, then released in an orbit that is calculated to balance its speed against the Earth's pull of gravity. Most satellites rely upon the forward rotation of Earth to slingshot them into space. (Earth rotates at more than 1,000 miles an hour, from west to east, taking its atmosphere with it.) If the satellites did not take advantage of this effect, they'd be fighting against it. For this reason, you'll never see a satellite in an east-west orbit.

Military satellites have other orbiting characteristics by which we can identify them. A distinctly north-south orbit allows them to look at the entire Earth as it spins below them, and yes, they can read an SOS spelled in flaming letters from 3,000 miles up. So if you see a north-south satellite, there's a chance that it's one of the three your GPS is talking to for coordinates (their signals will be scrambled during wartime). Unfortunately, the satellites that are of daily importance to the man in the street—the relays for telephones and television—cannot be seen. Not only are they 25,000 miles away, but they are geosynchronous, which means that they orbit at the same rate at which the Earth turns and cannot be seen moving among the stars behind them.

In some ways, satellites are easier to see than shooting stars. They're much less brilliant, but to find them requires only that you look for movement. They can be dependably seen anytime;

A satellite streaks across the field of this time-lapse photograph.

over fifteen may glide overhead in an hour. Their apparent size and brightness vary, and if not for their motion, they'd appear identical to the stars. Once spotted, they'll tinker along on a steady, quiet path until they disappear behind Earth or its shadow. They can best be found directly overhead and will cross the sky in about five minutes. They may frequently disappear then reappear—these car-size spacecraft tumble through space, visible only as they reflect the light of the sun. For this reason, it is best to look for satellites for a few hours before sunrise and after sunset. Otherwise they will have disappeared behind Earth's shadow and can no longer catch the sun. For the same reason, the summertime allows for the best viewing, as the tilt of the Earth puts us within less shadow.

THE WANDERING GODS

The five planets closest to the Earth—Mercury, Venus, Mars, Jupiter, and Saturn—are the farthest objects we can see moving in the sky with the naked eye. It is these five planets that have been the font of fortune-telling since 3,000 B.C.; No one could see the other three—Uranus, Neptune, and Pluto.

Like comets, man-made satellites, and Earth, planets are lit only by the reflected light of the sun and are its tiny satellites. Planets differ in appearance from stars mainly because they're so much closer; they're brighter and don't twinkle. Their brightness varies over time, though. They're *so* much closer than the stars that their journey through our tiny solar system can cause their distance from us to change by as much as 500 percent. They also go through phases, just like the moon, further altering their brightness. This makes it difficult to identify planets by their apparent size. All the planets travel through the plane of the ecliptic.

It's helpful to know that Mercury and Venus, called the inferior planets, circle the sun *within* Earth's orbit. The rest, called the superior planets, are outside Earth's orbit. This means that the inferior planets can always be seen close to the sun (at sunset and sunrise). So if you see a planet in the morning or evening, close to where the sun is hiding, it's probably Venus or Mercury—most likely Venus, as Mercury is smaller and more distant.

If you see a steadily burning light high in the sky or in the middle of the night, it's one of our outer planets. You may be able to detect a reddish hue to one of the outer planets, which would suggest that it's Mars, named after the Roman god of war. If you have exceptionally good vision, particularly if you are very young, you may be able to identify Jupiter by the moons circling it.

On the next page is a chart with information about the visible, permanent objects in the sky. By itself, it's a jumble of numbers that is best avoided. Used while sighting specific objects, however, it can be interesting and instructive.

These dry statistics come to life if you use your imagination. Saturn and Jupiter spin so fast that their spheres are flattened. Because Mars is so small, its gravitational force is low. A person on Mars could jump almost three times as high as on Earth. Saturn is so far from the sun, with such a dizzying swirl of moonrises and moonsets, that it would be confusing to separate day from night.

AIRCRAFT

High-altitude aircraft are easily confused with satellites, since they may appear as a distant white light slowly moving across the sky. Unlike satellites, aircraft produce their own light. An object seen high overhead more than a few hours before sunrise or a few hours after sunset is an aircraft; at that time satellites lurk completely in the Earth's shadow. Aircraft can be visible for some minutes before the sound from their engines reaches you. A jet airplane's condensation trail, known as a contrail, is another dead giveaway, especially visible on a moonlit night. When airborne, the only lights visible on aircraft are the flashing green and red anticollision lights, but these can appear white at a distance (jets

Properties of Visible, Permanent Sky-Objects

	Length of Day	Length of Year	Distance to Sun	Diameter	Temperature	Moons
Earth	1	1	93,000,000 miles	7,500 miles	–100 to 100° F	1
Mercury	59 Earth days	0.24 Earth days	0.38 Earth distances	0.38 Earth diameters	–300 to 1000°F	0
Venus	243 Earth days	0.6 Earth days	0.72 Earth distances	0.95 Earth diameters	850°F	0
Mars	24 1/2 hours	1.9 Earth days	1.5 Earth distances	0.5 Earth diameters	–225 to 60°F	2
Jupiter	10 hours	1.2 Earth days	52 Earth distances	11 Earth diameters	–225°F	16
Saturn	10 1/2 hours	29 Earth days	95 Earth distances	9 Earth diameter	–300°F	Over 20
Moon	27.3 Earth days	NA	240,000 miles from Earth	0.27 Earth diameters	260°F (day side)	NA
Sun	25 Earth days	NA	NA	110 Earth diameters	11,000°F	NA
Milky Way galaxy	Rotates every 250 million years	NA	NA	586,800,000,000,000,000 miles	?	NA

The first four planets, Mercury, Venus, Earth, and Mars, are largely solid bodies. Jupiter and Saturn are liquid and gaseous.

cruise at an altitude of 6 miles). The only time an aircraft actually shines a white light is during its lengthy ascent and descent.

It's fun to survey the sky and recognize a glow above the horizon as a familiar city or town. If you're viewing from a remote country location, however, you will often see patches or bands of light that are not of earthly origin.

AURORAS

Auroras appear as immense, billowing curtains of light that can span the entire northern sky and last for hours. If music could be written in the sky, this would be it. And yet this ghostly play of light, at times faint, at other times explosively brilliant, is soundless. Rarely is any astronomical phenomenon more breathtaking than the auroras. Unfortunately, they are unpredictable and increasingly rare outside the northernmost states, though they have occurred as far south as Texas.

Vic Winter

An Aurora, Powell Observatory, Kansas.

Cold nights and high latitudes are a helpful prescription for auroras; another is sunspot activity. Why? Auroras largely occur in conjunction with the very regular patterns of sunspots that occur in eleven-year cycles on the sun. These are intense regions of magnetism breaking through the surface, the breeding grounds for the violent solar flares that burst from the sun and shoot into the solar system. When this electromagnetic radiation streams into the Earth's magnetic field, auroras are the result. So, along with shooting stars, this is one of the few part-time members of the sky that produces its own light. Interestingly, scientists in the Arctic have found patterns in the auroral displays that tell them the time just like a clock.

ZODIACAL LIGHT

Another hazy glow, much more common, can be seen from any darker location on most nights. A quick check of your bearings will reveal that this is where the sun had set. Don't be fooled—this is not a distant city, an aurora, or the dying light of the sun. This vaguely triangular pillar of light that makes it difficult to see the stars in that region is known as the zodiacal (zoh-dye'-uh-kul) light, and it may reach as far as 50 degrees into the sky.

This is merely the rays of the long-departed sun reflecting off the interstellar dust in the plane of the ecliptic, which is also the pathway of the constellations of the zodiac. The astrological signs, such as Aries, Leo, and Virgo, are named after these constellations and were chosen because the sun travels through them, and not for any other reason.

The zodiacal light is best seen an hour or so before sunrise and after sunset. It is quite different from the Milky Way. Though both are long, hazy patches of light, the Milky Way enriches the sky with its winding river of stars, whereas the zodiacal light obscures it.

UNIDENTIFIED FLYING OBJECTS

Every one of us has seen a UFO—an unidentified flying object. If the average person can't identify a planet, then obviously there's a lot more up there he or she can't identify. But there's a big difference between a UFO and a spacecraft carrying extraterrestrial visitors.

Many scientists believe that there must be life—intelligent life—elsewhere in the universe. At the same time, few would agree that we've actually been visited by that life. And this is the crux of the UFO controversy: no one denies that there are strange things to be seen in the sky, and most believe intelligent life could exist elsewhere. The question is whether that life has come here.

The federal government tried to solve the mystery in the 1960s and '70s with Project Blue Book, an official investigation into UFOs. It became uselessly bogged down in debate. The incredible explanations put forth by the Air Force even today concerning Roswell, New Mexico, suggest that the topic is still too explosive.

Clearly, strange things have been seen in the sky and reported by credible people all over the world. The Bible, in the books of Daniel, Ezekiel, and Revelation, describes fantastic objects in the sky. Unidentified flying objects have been reported in the United States since the 1880s and '90s, when many people claimed to see large, blimplike objects. During World War II, pilots on both sides sighted strange fireballs, termed "foo fighters" by the Allies, accompanying their aircraft. Each assumed it was a secret weapon of the other.

The term "flying saucer" came about in 1947 when businessman Kenneth Arnold was flying over Washington state and reported seeing a row of silver disks dipping up and down. Since then, several cycles of UFO sightings have occurred, most of which were ultimately downgraded into IFOs—identified flying objects. People often mistake as UFOs aircraft, planets (especially Venus), and stars close to the horizon that are taking on unusual colors, just as does the setting sun. Even the most fanatical UFO monitoring groups will agree that the vast majority of cases can be explained away.

Unlike every other object described in this chapter, there's no best time or place to watch for UFOs (otherwise they wouldn't be UFOs, would they?). If you spend any amount of time looking up, however, you'll see one—an object that you don't have the ability to identify. There are several criteria to distinguish truly strange lights from anything else in this chapter and from any conventional explanation: if an object makes a sudden change in direction, as though it were ricocheting; if lights come together and stay together, or if a light or lights were together and then come apart; and if your location is nowhere near a military installation. This last is simply because military aircraft don't always abide by the rules governing civilian aircraft for height, lights, and so on. And the military has equipment—helicopters and vertical takeoff and landing craft—that can perform remarkable maneuvers.

The verdict is still out as to whether we have actually had visitors from another world. There is no conclusive evidence to demonstrate that it has happened, nor is there any to prove that it could not. Although many scientists think it is possible, they wonder why visitors advanced enough to travel across the universe are so reluctant to make their presence known—or why they couldn't remain completely unknown.

SUMMARY

- It's extremely likely that you've seen all of these "wandering" lights before, without even knowing it. All but comets and auroras are common. But if you don't know what you're looking at, it's hard to be sure of what you're seeing.
- A simple but important skill is to learn to trust what your eyes tell you. Do you see color in that light? Is that

object moving slowly or is it really stationary? Is the sky a little brighter in that area? It takes a little time to train and trust the eyes for what they see. Remember: if it moves, it's not a star, and it's within our solar system.

- Meteors, nicknamed "shooting stars", are streaks of light that last for a few seconds or less. Usually several can be seen at any hour of the night, anywhere.
- Satellites look like moving stars or high-altitude aircraft. They take several minutes to cross the sky, and several may be seen each hour. They can best be seen an hour or two after sunset or before sunrise.
- Comets are rare and resemble a fuzzy patch of light, brighter in the nucleus. They take several months to move across the sky, and their appearance is widely publicized in the media.
- Planets appear as bright stars that don't twinkle. Their motion can be detected only over a period of weeks or months, and they travel in the same path across the night sky as the moon or the sun does during the day. They may be seen anytime.
- Aircraft have blinking red and green lights, usually will make noise after a few moments, and may leave a vapor trail.
- Auroras are vast curtains of light that appear to move. They are generally visible toward the north in the more northerly latitudes.
- The zodiacal light is a roughly triangular tower of light that will appear an hour or two before sunrise and after sunset.
- If you don't know what something is, it's a UFO. Other than seeing bug-eyed freaks emerge from a spaceship, the only criterion that would clearly set off something as unusual would be an object that makes sudden changes in direction or composition.

THE USEFUL UNIVERSE

Most of us cannot imagine the distant stars as being of practical value to anyone but ancient mariners and nuclear physicists. Yes, the stars are pretty to look at, and our ancestors had some fascinating ideas about the sky. But the thought that they may actually be useful to anyone today not peering through a telescope may seem strange. This chapter is about the ways that we can really *use* the sky, not just look at it and have deep thoughts. Granted, the practical application of these methods is most useful to the outdoorsman—the hiker, sailor, or hunter. But the knowledge underlying these methods is so fundamental to our understanding of the night sky that one could trample it in our hurry to get to the rest. Learn by doing they say and there is no more practical way to learn about the stars than to actually use them.

It was the unfailing dependability of the stars in the sky that allowed mankind to emerge from dark caves and build empires. This dependability is just as valuable to us today in discovering the most essential of knowledges: time, location, and direction. So why not bring a watch, a map, and a compass? Certainly the outdoorsman should have all three at all times. But how many of us know how to use a compass or read a map? Man-made tools require a formal and contrived knowledge; many, many people have found it better to rely on more natural means. There are three that come instinctively:

- Landmark recognition: "We saw that before, coming from the other direction."
- The inner clock: "I woke at seven on the dot, without an alarm."
- Common sense: "We've been driving east into the sun, and she said she'd be on the north corner."

This is how we think and behave every day, and it's not difficult to expand what we've been doing every day of our lives into the night.

Watches and compasses can be lost or damaged, and backup is critical. Worst of all is having an instrument that may be working improperly but there's no way of telling—the compass that seems off but is all a hiker has to find his or her way out. This is a fear that grips every beginning outdoorsman.

At such times, it may be better not to have a compass at all. Certainly this is so if you know how to read the signs of the sky. At the very least, knowing how to check your instruments against the original source comes in handy when you're 3 miles from the nearest road and don't know which way to head.

But don't expect to plow through this chapter and then stomp out the door into the wild blue yonder. It will take time to become familiar with each technique, and each requires practice in the field to master it. Then, too, the necessary skymarks will not always be present. Work with what is available, one step at a time. Remember, everything you learn in this chapter will only serve to reinforce what is in all the others.

TELLING DIRECTION

By night, the simplest way to tell direction is to look at the stars just barely above the horizon. You may have to search a few moments to find a prominent star. To the east, they will be rising.

In the west, they will be setting. In the south, they can be seen to move in a flattened, downward-facing arc. This can be observed in only a few minutes. It's just the opposite in the north, but it's harder to observe the increasing lack of motion that characterizes the stars as you look northward (a clue that you're looking north), and the North Star will not move at all.

The best method for learning direction is the one made popular by the Arabs five hundred years ago: finding the North Star, which resides extremely close to true north. This is only by happy coincidence, not because of any special affinity between it and the Earth. (It deeply impressed our ancestors to see a magnetized object point toward it—no wonder they believed in the influence of the stars.) There is no corresponding star in the Southern Hemisphere, and we are extremely lucky to have this one. This star-marked point is directly above the Earth's axis and is more accurate than your compass. It's so dead-on, as a matter of fact, that it can be used to make sure your compass hasn't become demagnetized.

To locate the North Star, first find the "pointer stars" in the constellation the Big Dipper. It's shaped like a squarish soup ladle (it really looks like its name) and may be found in any position—upside down, level, on end—in the north. It's roughly 25 degrees long, and all of its stars are about the same brightness, which is why it's so easy to find. Using the two stars on the outer edge of the bowl, draw a line out of the bowl. This line will shoot through the North Star, roughly four fist widths above the horizon.

If the Big Dipper and North Star still elude you, you can cheat and use a compass your first time out—but just this once. Follow the red half of the arrow to the horizon, then count up roughly four fist widths. The farther south in the United States you are, the closer it will be to three fists; the farther north, the closer to five fists. This may give you some indication how the North Star has been used to determine latitude, or your position between the equator and the North Pole. Since the North Star never moves, if you move up or down on the face of the Earth, its position in the sky will appear to go up or down. This is how it's been used by travelers for thousands of years.

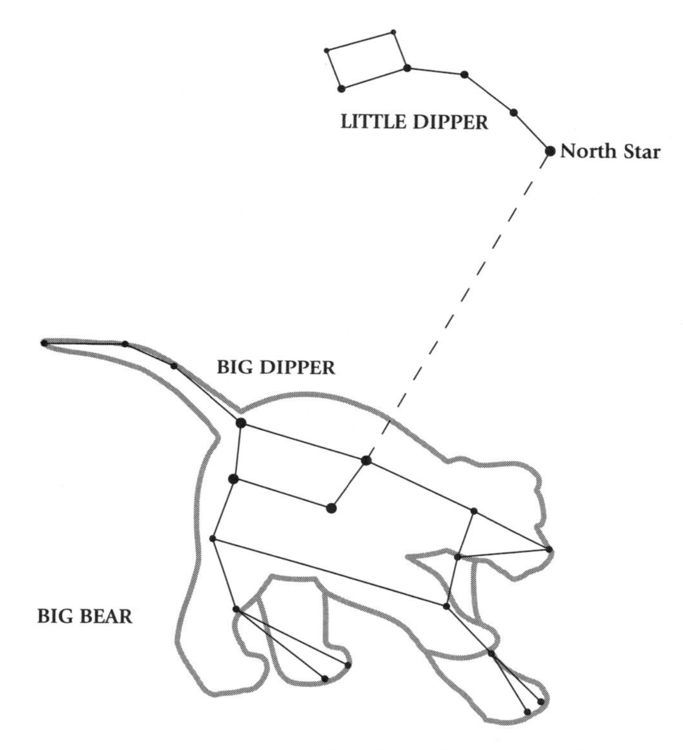

Approximate north can also be found by using the stars in the bold winter constellation Orion the Hunter. (page 64) A line drawn from the star Saiph (pronounced sife), in the hunter's left foot, through shimmering red Betelgeuse (bee'-tel-jooz), in his left shoulder, will point approximately north. This particular piece of information is not very useful by itself and may be distracting for the beginner. But together with other direction-finding tools, it will help you develop an instinct for finding your way in the outdoors.

Here in North America, we are well above the equator and are effectively looking down on the plane of the solar system. This is why the path of the sun (the plane of the ecliptic), moon, and all of the planets is always in the south. Their positions on this pathway may range from extreme southeast to extreme southwest, but they're in the south nonetheless. This gives us several direction finders.

Our position in North America allows us to look down on the plane of the solar system, so the sun, moon, and planets are all seen in the south. If we lived in South America, we'd see them all in the north.

The morning shadows of the sun will point to the northwest. They will then incline to the northeast through the afternoon. These shadows change so slowly that once their angle to your direction of travel has been established, your course need not be checked again for over an hour. Using the sun to find your way is surprisingly useful, especially to a city dweller. The moon can be used in a similar fashion at night. It is helpful to remember that the sun and the moon always move to the right as you face them, from east to west. Because of the southerly position of the plane of the ecliptic, we also know the following:

- The sun is at true south at high noon, and its shadow points toward true north.
- The full moon at its highest point (about midnight) is at true south.
- The quarter moon (what appears as a semicircle) around sunrise or sunset is in the south.

The sun appears to move east to west, so it follows that if you could somehow plot its course, you'd have an east-west line. Drive a stick into the ground or find a shadow of a branch. Mark the tip of the shadow. Half an hour later, mark it again. A line drawn from the first shadow tip through the new one will run approximately east-west. Since the sun always moves to the right, or west, the motion of the sun will tell you which end is which. A perpendicular line through this one will run north-south.

If it's a cloudy day, open a knife and hold it horizontally over a light, even surface, such as a book or a flat rock, with the blade turned vertically as though you were slicing. (page 39) Turn it until the shadow cast by its edge is narrowest, or the shadow is the same width on either side of the blade. This indicates the direction of the sun. Just because it's a cloudy day doesn't change the fact that Earth's source of light radiates from a single point. Mark each end of the knife on a flat surface. An hour later, place the butt of the knife where it was before, and pivot the knife until once again the shadow of the blade is narrowest. A line draw from tip to tip will go east-west.

Generally speaking, approximate directions are good enough to get you home. Still, it is a fair question: Who can afford to spend an hour waiting for a shadow to move or the sun to peak, only to get your direction once? It'll just have to be found all over again in a little while, won't it?

The fact is, your travel direction only needs to be found one time. No matter how hopelessly lost you may be, once you've found your line of travel, it's easy to keep track of it. Simply sight on a distant object along the line of travel and note a landmark where you are now. When you get to the distant object, sight back to your original starting point. A line through this and the one-time distant object (your new location) will allow you to continue along your original bearing.

CHECKING YOUR COMPASS

If you ever become lost while hiking, you may wonder if your compass is off. Using the sky is the best possible means to check that your compass didn't become demagnetized.

Try to find the North Star using your compass. You can expect it to point a little askew of this point because of Earth's magnetism. The magnetic field that surrounds the Earth flows along a north-south orientation. (Solar flares radiating into this magnetic field are the cause of auroras.) This huge internal magnet pulls on your compass needle.

But this field does not circle the globe as uniformly as the lines of longitude spindling around it, and it has many distortions. Luckily, these distortions are fairly regular and have been plotted everywhere. Most maps, except for common highway maps, show a location's magnetic deviation, known as magnetic north, beside or within the map's north-south indicator, known as the compass rose. Any tug on the needle beyond this magnetic

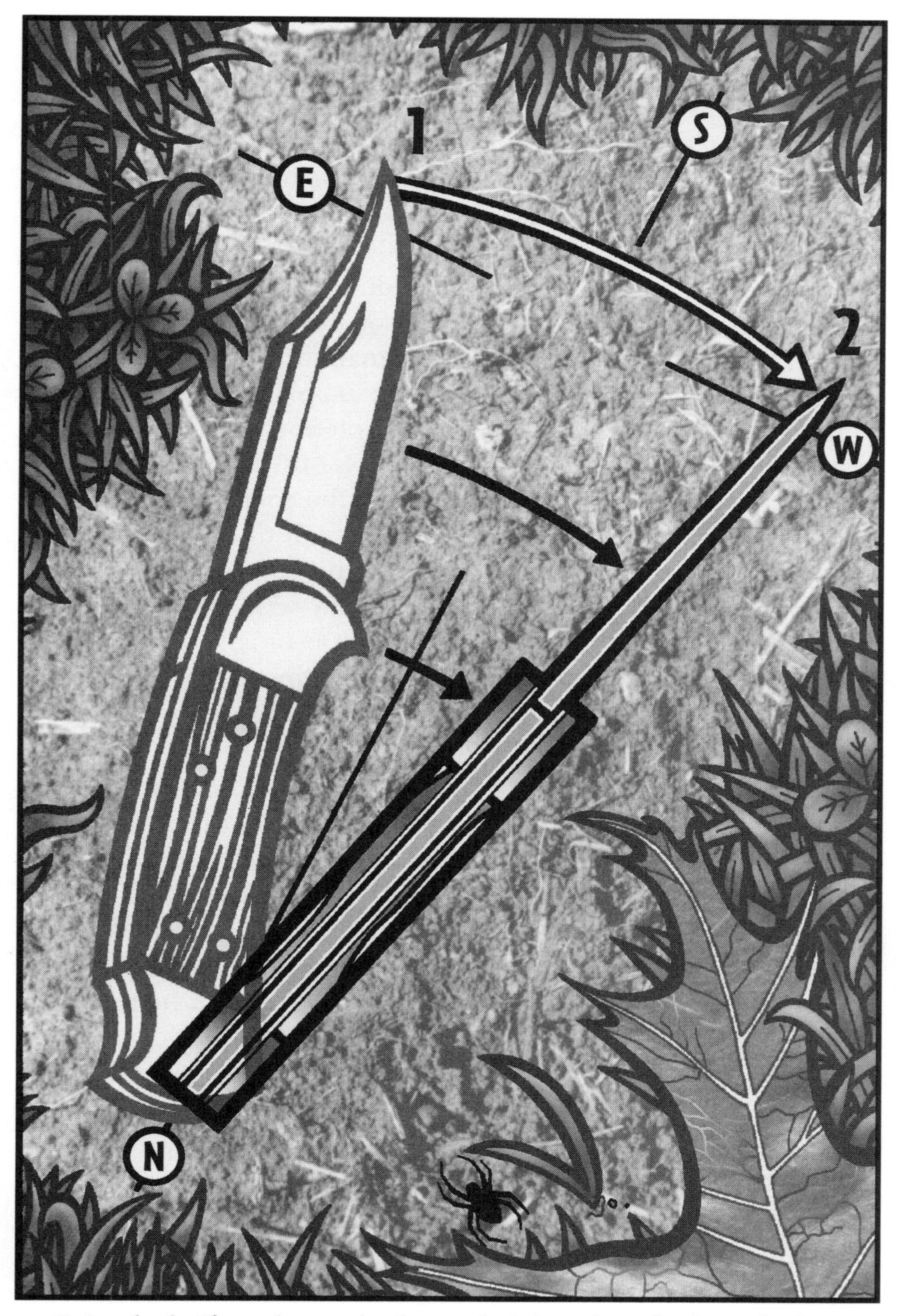

Point the knife at the sun in the south (1) and mark where the butt and point are. A little time later, pivot it to follow the sun. A line from 1 to 2 will run east-west.

north is a measure of how true your compass is. A few degrees of deviation won't get you into any trouble.

It's better to use natural direction finders, however. If you can find your direction in the sky, it's easier to hike without a compass, which requires that you constantly check and recheck your line of travel and allow for magnetic deviation. Using a map to plot your course is simpler as well, for you only need to know true north. Draw a north-south line on the map through your present location, using the north-south of the map's compass rose to guide you. Now draw a line from your present location to your destination. This is your line of travel. Point the north-south line at the North Star (or at the midday sun in the South).

Now look for a landmark on the horizon or star in the sky that the line of travel points to. This is the beacon you will follow. If you're using a star to guide you, you may wish to line up the map again with the North Star after an hour or so to see if your guide star has moved. If you're using a landmark, this won't be necessary. If you are in a deep woods that obscures your landmark, use the sun's shadows.

It's easier to travel with the stars by night than the sun by day. Who would want to travel at night? Many people prefer to, especially in the desert. The heat is down, the animals are out, the wind has died, and it's a beautiful time to look at the sky.

Night hiking is a wonderful way to experience the outdoors. You will learn that the darkness is not to be feared; there is a whole new world to explore. You will learn to listen better, use your eyes differently, and enjoy a sense of concealed yet confident activity. It may be a good idea to wave a stick in front of your head at times or use the "tiger step," in which you make a sweeping crescent step with the advancing leg before shifting your weight from the rear. Bring a flashlight just in case, though you'll find you can see better without it. As you hike quietly in the night, you will naturally begin to search for whatever light is available, and you'll begin to understand why men and women have always looked to the stars.

By aligning 'north' on your map with the North Star, you can confidently plot your course.

TELLING TIME

By themselves, any one of these time-telling techniques might be helpful for telling the time only once, at a moment that is unlikely to serve your needs. Working together, they overlap and reinforce each other so that it is possible to have a reasonably good idea of the time throughout the day and into the night. How nice it would be to live in a world where a reasonably good idea of the time is the biggest hurry we'd ever be in.

Believe it or not, the verdigris sundial that stands so proudly in Aunt Tillie's flower plot works just as well on a rainy day as on a sunny one. This is because unless you have access to complicated schedules and positional information, it doesn't work at all. It's useful only in determining the position of the sun, which can be gauged rather more accurately by looking at the sun itself.

Even high noon on the dial is unlikely to match 12:00 on your watch, owing to daylight saving time and your distance from your time zone's boundary. Suppose one location is 300 miles west of another, but both are within the Mountain Standard Time Zone. Though both have the same official time, the sun will be overhead at each place at different times. Just because people divide the Earth into full-hour designations doesn't mean nature does so.

Man's artificial (but useful) manipulations of the time make it impossible to look at the sky and say, "It's 6:17, Bob, turn the steaks." Using solely natural means is most useful in judging how much time has *passed*, not exactly what time it is. On the other hand, if you know what the local time is when you start using astronomical signs, the sky is a reliable chronometer for both local and natural time.

Because the Earth revolves around the sun, we see each individual star rising and setting four minutes earlier every night. Thus, if you know the time that a given star rose above the horizon, you may dependably subtract four minutes for each day afterward and know the time. Say you saw the star Arcturus rise at 7:44 on Tuesday evening. Then suddenly your watch batteries died! A week later you would know it was 7:16 when you saw it rise again (seven days times four minutes subtracted from 7:44). This is an excellent method for checking the dependability of your watch.

Unfortunately, the sun is so close to the Earth that it cannot reliably be used for so exacting a purpose. Early men initially used the passage of the sun to establish the length of the day and year. In just a few years, however, this would always lead to inconsistencies that had devastating results.

It's precisely because the stars are so distant that they are so dependable. We cannot see their actual motion, so it is only the motion of the Earth that makes them appear to move, and this we can allow for. For this reason, we've come to adopt the sidereal (star) year over the solar (sun) year. It is our position relative to the stars, not the sun, that is now used to measure the year—just as you should set your watch.

But knowing the time only at sunrise and sunset is almost as useful as a broken watch, which is also accurate only twice a day. There are several methods that, used together, can keep you on time throughout the day, without a man-made timepiece.

If the moon is present, we may count on its chronic tardiness to tell the time. As well as turning with the stars, it revolves around the Earth. So every night—and day—it delays its appearance onstage by about fifty minutes. To tell where the moon is today compared with yesterday, sight it over a feature such as a tree, saddle, or mountain. When it crosses that same point the next day, it's roughly fifty minutes later than the day before.

This is the same motion that causes the moon to move across the background of stars at the speed of one moon width every hour, or half a degree. But it's difficult to judge this, if only because the surrounding stars are overwhelmed by its brightness.

It's best, then, to find two stars that are endpoints for a line running next to the moon, and judge its passage compared to that.

What if there's a new moon—that is to say, no moon at all? This occurs when the moon is directly between the sun and Earth, and all we can see is its shadow, or nothing. Never fear, for the night sky is the very model for our clocks and watches. Remember how a line drawn through the bowl of the Big Dipper points toward the North? The Big Dipper circles the North Star just as the hour hand circles a clock, and so does this stellar line sweep out the passage of time. The only difference is that the celestial timepiece is on a twenty-four-hour dial, not a mere twelve, and it moves counterclockwise. The same turning of the Earth that inspired man to mark time with a round-faced clock is the same motion that makes the starry sky appear to turn as well.

It's too much to expect, of course, that when our clocks point to 3:00 P.M., the stellar clock would too. But the motion is the same. Suppose you went to bed and saw that the pointer stars pointed straight down, at a watch's 6:00. When you woke up they had moved counterclockwise, to 2:00. The four "regular" hours are doubled—eight hours have passed since you went to bed, and the sun will be up shortly.

If you have an idea what time the sun has been rising and setting (check the weather section of the newspaper or observe yourself), you know how many hours of daylight there are. So if the sun is a quarter of the way across the sky, simply add a quarter of the hours in the day to the time you know that the sun rose. This will give you a useful, if approximate, time.

CHECKING YOUR EYESIGHT

Ever wonder how good your vision is? Look to the Pleiades, or Seven Sisters (directions for finding them are in chapter 5). If you can count seven stars, your vision is average. Twelve is exceptionally good. If you can see that Vega, in Lyra (page 67), is a double star, this is also proof of average vision.

The motion of the night sky—the model for our own watches and clocks.

HOW BIG ARE THE SEAS?

When you visit the sea, you may notice that the height and fury of the tide can differ from one week to the next, even though there have been no storms. Look far, far away for the cause—to the sun and moon.

To begin with, the moon is responsible for the twice-daily tides. As Earth's gravity clutches the moon close beside her, so, too, does the moon try to hug the Earth. But being so much smaller, all it can do is coax the seas from their beds, swelling them to break higher against the shore.

This power varies. Roughly twice a month, during the new (invisible) and full moons, when the moon is in line with the sun, directly behind or in front of the Earth, the gravitational pull of both draws on the oceans. These combined forces create the spring tide.

Neap tide occurs roughly twice a month as well. This is during the quarter moons (semicircle), when the moon is at right angles to a line drawn between the Earth and sun. Here the sun-moon

Use the star cluster Pleiades to check your eyesight.

The moon—the cosmic controller of our tides.

forces tend to cancel out. This contrast can be seen in just a week's time. There is no better demonstration that the greatest power on Earth is in the sky.

I'm not going to pretend that we have now arrived at a more natural and thus better way to tell time, direction, and location; there is simply no substitute for a compass, watch, and map. But to rely on them solely is like driving without a spare tire. What if the compass becomes demagnetized, you lose your watch, or *you* become lost? These techniques cannot be easily mastered in one trial; they take practice. But remember, as a child, how long it took to learn to read a clock.

There is a sixth sense that transcends taste, touch, smell, sound, and sight. This sense enables us to move with ease and confidence in an unfamiliar and unforgiving environment. That sense is, quite simply, intelligence, the common sense born of experience. Continually checking and fumbling with specialized instruments forces us to focus on technology, not the place, and suggests a certain sense of not belonging. And who needs to know that it's 11:37 when you're trying to relax in the outdoors? That's one place where you should learn to steer by your own lights, the lights of the stars.

It's not merely ego or gee-whiz tricksterism that pushes one to learn to instinctively know the way or time of day. It's too much work. It takes skill, patience, and a passionate desire to understand the world around you. The sky above is more resolutely set apart from our existence than anything around us, yet nothing else is so importantly shared by every creature on the planet. Many times I have checked the compass to find north, only to forget it moments later. But when I find it in the sky, whether by the north of Polaris or the south of the sun, forever and unchanging above our world, I never forget.

THE GREAT LABORATORY

Sir Kenneth Clark begins his book *Civilization* with the simple assertion that the basis of civilization is confidence. Early peoples had precious little to be confident about. There was nothing certain about the weather, abundance of game, water, or the harvest. Their neighbors might drive them out or kill them. Even the realm of the gods was unsettled. The moon changed shape, the planets dodged in and out, the sun might disappear, and the entire sky with its ten thousand stars whirled and pitched.

We like to picture tribal peoples ending the day gathered around a glowing fire. Yet this fails to take into account scarcity of wood or the need for security. If our ancestors looked to fire for light and comfort at day's end, it wasn't on the ground before them; they turned to the darkness, to the lights burning a hundred trillion miles away. Seeing how the great sun's heat and light influenced the Earth, they naturally looked to the lesser moon and planets for influence in their own lives.

In doing so, they couldn't help but see patterns in the stars. Not just pictures of monsters and heroes, but patterns in timing and position. When these changed, they did so with reassuring coincidence with the life cycles of plants and animals and the round of seasons. And they did so dependably. When people finally acquired the courage to contrive gods loftier than the beasts around them, they found them in the lights above.

The life of the hunter-gatherer was meager, brutal, and short. It was impossible to keep the band together as game and vegetation dwindled. The times people spent together in peaceful abundance were special largely because they were so rare. The stability of an agricultural existence came only to those who observed both the life cycles of vegetation and the patterns of the stars. The calendar of the sky was indispensable in deciding the precise times to sow and cultivate. When the Earth journeyed around the sun and brought key stars over the horizon, these were the signal lights to commence planting. This was the beginning of civilization, that stable village life that many think was the best of all possible worlds.

Because people's lives were so dramatically improved by agriculture, they began to worship the stars that made it possible. And this reverence was terrifically abused by king-priests who controlled the knowledge of the heavens. Their ability to predict the behavior of the stars, sun, moon, and planets was critical to maintaining control. So long as they proved their ability to know the will of the gods, they were free to attach to it any colorful mythology they wished: the Earth sits on top of four elephants on top of a turtle on top of a snake; the youth must renew the fertility of Mother Earth with fantastic sexual activity; women are forbidden to look on the face of the king. Because the priests appeared to control the most powerful forces on Earth, the faithful masses could only trust that these fanciful notions were true as well.

As deceptive as this was, it was just this focus of human endeavor that resulted in some of civilization's greatest achievements: the ziggurats of Egypt and South America, the stone rings of Britain, the observatories of Arabia and India. The knowledge required to erect these monuments—geometry, trigonometry, and mechanics—became the building blocks of science.

Yet the very success of agricultural life forced people to desperate measures. Around the world, populations grew beyond the capacity of the land to support them. Wars forced the losers to

In ancient times, those who could forecast the behavior of the heavens controlled the people.

migrate and encouraged the conquerors to expand further. For thousands of years, these expansions were made possible and then limited by local methods of navigation: the flight paths of birds, wind directions, and the colors and patterns of waves.

Then man peered into the darkness and found a better way. By noting the position of the sun and the North Star, it was possible to determine one's latitude. This meant that the navigator of a caravan or ship always knew how far north or south it was. This enabled the ancient Phoenicians to sail from the Eastern Mediterranean to Britain.

Two thousand years later, the Vikings journeyed from Finland to Labrador. Simply by knowing how far north or south they were and reckoning the distance traveled, they could guess what lay to

Vic Winter

Simply by watching the turn of the stars above, early sailors could take a simple reed boat like this across the Pacific or Atlantic.

the east or west. It was now possible to cross the seas and oceans. If man had not looked to the heavens, the five hundred years that have passed since the discovery of the Americas would have passed precisely as the two thousand before it.

The Arabs discovered that individual stars could be used to steer toward specific locations. Along with the compass, this made possible the Great Age of Discovery. Western civilization, once confined largely to the Mediterranean and Europe, exploded onto every continent within just a few hundred years. Commerce followed discovery, and conquest followed commerce. The material wealth of Central and South America poured into Spain. The wealth of expropriated labor from North America, Africa, and Asia poured into the rest of Europe.

The disastrous effects of these conquests on the peoples of the nonwestern world were matched by extraordinary changes in the West. Up until now, astronomical phenomena had been used to reinforce religious beliefs; they were now used to repudiate them. The same starry-eyed adventuring that had made the Christian nations so wealthy and powerful began to undercut their very faith.

The Church had taught that the Earth was the center of a tiny spherical universe that revolved around it. All heavenly bodies were created perfectly by God and moved in perfect circles; so was the rest of the universe perfect and unchanging. All experiences and observations were made to fit into this physics-cum-theology that was the Church's claim to power. This process is known as *inductive reasoning*.

In 1543, Copernicus published *De Revolutionibus Orbium Coelestium (About the Revolutions of Heavenly Spheres)*. This was the first time anyone had formally suggested that the Earth revolved around the sun. Even though it was based on some sketchy science, it explained the motions of the solar system more satisfactorily than the physics of Aristotle, the Church's torchbearer.

A sun-centered solar system and the science that grew up around it answered important questions, such as why the planets

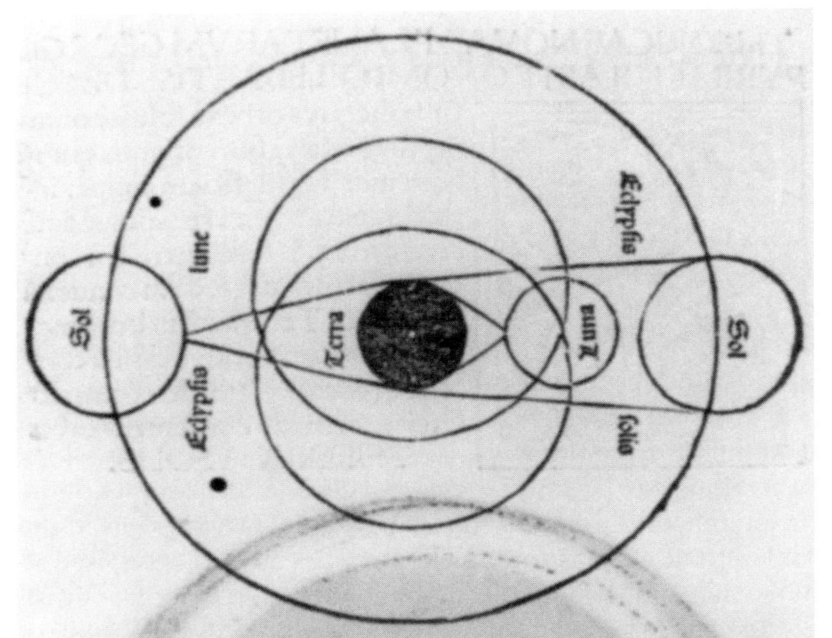

corp⁹ solare. Vnde obũbrabit nobis claritatẽ sol:& ita sol patiet
eclypsim:nõ qa deficiat lumie:sꝫ deficit nobis,ꝑpter interposi/
tionẽ lunę iter aspectũ nostrũ & solẽ.Ex his pꝫ ꝙ nõ semp̄ ẽ ecly
psis solis ĩ cõiũctiõe siue ĩ nouilunio. Notandũ etiã ꝙ qñ est
eclypsis lunę:est eclypsis ĩ oĩ terra:sꝫ qñ ẽ eclypsis solis nequaq̃ꝫ
ĩmo ĩ uno climate ẽ eclypsis solis:& ĩ alio nõ.qđ cõtingit,ꝑpter
diuersitatẽ aspect⁹in diuersis climatib⁹. Vnde Virgili⁹ elegantis
sime naturas utriusqꝫ eclypsis sub cõpendio tetigit dicẽs. Defe
ctus lunę uarios solisqꝫ labores.Ex p̄dictis pꝫ ꝙ cũ eclypsis solis
eẽt in passiõe dñi:& eadẽ passio esset in plenilunio:illa eclypsis
solis nõ fuit natural:imo miraculosa cõtraria naturę:qa eclypsis
solis ĩ nouilunio uel circa debet cõtingere.Propter quod legit
Dionysiũ ariopagitã ĩ eadẽ passiõe dixisse. Aut deus naturę pa
tit:aut mũdi machina dissoluet. Opusculũ sphęricũ Iohãnis
de sacro busto explicitum est.

3 2

Irrefutable proof that the sun does indeed go around the Earth.

appear to move backward at times and why the stars' positions don't change when one travels, as the sun's does.

But Copernicus's fear of the Church led him to publish his book only on his deathbed. Even then, the book's preface suggested that it was only a fanciful notion. But his ideas caused such a stir at the time that even now we use the world *revolution* to describe an event that threatens to overturn everything else.

A hundred years later, Galileo Galilei turned his primitive telescope skyward. What he saw changed the world forever: a moon with a cratered, imperfect surface; Venus changing its shape (in phases like the moon); spots on the sun; Jupiter with four moons, proving that the Earth was not the center of the universe. Galileo was hardly the first to see these objects; he was simply the first to formally describe them.

Anybody could look into a telescope and see for himself that Galileo's observations were true—this was no mere hypothesis, as with Copernicus. The Church's response was to threaten Galileo with imprisonment and torture; naturally, he recanted. But his conclusions were so obvious to anyone with a simple spyglass that once his observations had been publicly stated, a sea-change of opinion began.

The demands of warfare, commerce, and navigation—pressures that would overwhelm the power of the Church if ignored—forced the Church to adapt its beliefs to reality. This process is known as *deductive reasoning*.

Building on the painstaking observations of Tycho Brahe, the astronomer Johannes Kepler established the exact motions of the planets. This paved the way for Sir Isaac Newton's discovery of the three laws of physics in 1687: an object in motion tends to remain in motion; an object will accelerate in proportion to the force exerted on it; and for every action there is an equal and opposite reaction.

It may seem strange that the discovery of the most fundamental laws governing our physical existence was done by observing tiny dots of light in the sky trillions of miles away. But just as the sky provided the Earth's only dependable clock, calendar, and compass, so, too, was it the purest proving grounds for testing the laws of nature. The laws of physics control the jump of the tiniest flea on Earth, but those laws could be observed and measured only in the severe purity of outer space.

Unfortunately, there was a long delay between such scientific discoveries and the means to apply them. But Newton's explanation of why an apple fell from a tree was so compelling, and the deductive method so powerful, that they were applied to every aspect of life: the development of machines and engines, the construction of larger and more efficient buildings and ships, the mapping of the globe—even ideas of social justice, human evolution, and economics. A whole truckload of new laws followed. This was the Age of Enlightenment.

The Age of Discovery had been characterized by a new diversity and abundance of material things, along with the question "Who knows the will of God?" But the driving force behind these changes had been greed—greed for new lands, slaves, spices, and gold.

Rapid, radical change characterized the Age of Enlightenment—changes in opportunity, travel, communications, the length of life, and individual wealth, as well as the ability to provide heat in the winter and light at night. Every societal structure and belief became subject to debate, and the question arose "Is there a God?" Greed had not lost its driving force, but now scientific insight and social creativity led the way for change. This new way of thinking demanded that decisions be based on observation and experimentation, not merely theories that were agreeable to those in power. A revolution began that swept over every aspect of life, in large part because of Galileo and his telescope.

But Newton's laws were not sufficient to explain reality as people knew it, though many, even today, would have preferred to leave it at that. Enough change, enough progress! Scientists found that Newton's laws needed refinement, as they could not answer simple, crucial questions: Why doesn't the sun burn itself out?

Why is the sound of a train different for a passenger than a person watching it go by? Why does the acceleration of a car feel like the force of gravity holding us to Earth? These obscure questions were central to Einstein's general theory of relativity.

To prove his theory, Einstein needed to observe the effect of gravity on passing objects. Again, only an especially pure, extreme environment could test the theory, an environment that couldn't be found on the Earth. Einstein needed to watch the ability of the sun's gravity to bend the light of a star. But this was not enough: the sun's own light would obscure that of the star. Only during an eclipse, when the moon darkened the sun's rays, could the star's light be seen. Two expeditions were sent to view the eclipse, one to Brazil, the other to Principe, off the coast of West Africa. They confirmed that light could be altered by gravity.

Einstein's theory can be summarized thus: the acceleration caused by a gravitational field (such as Earth's holding us on its surface) cannot be distinguished from the acceleration caused by motion (such as is felt when a car moves from a stop).

This simple theory was responsible for the development of the atomic bomb and nuclear power. (Believe it or not!) It brought blessings of a less mixed nature as well: nuclear medicine, transistors, space exploration, and satellite communications. Unfortunately, the Nuclear Age is far too young to have fully realized its benefits. Certainly it has confirmed our confusion over the nature of reality itself, and it threatens an end to human existence as never before.

Scientists say that a successful experiment is not one that gives them the results they wanted, but one that was conducted properly, where they could be confident that all conditions had been controlled and allowed for. For this simple reason, men have tried to find answers in the stars. The stars have been responsible for man's advancement again and again. But no matter how ideal the laboratory conditions have been, it is fair to ask, "Was the experiment conducted properly? Did we get the results we were looking for? Had we any business conducting such an experiment?"

Einstein's general theory of relativity introduced the blessing of amazing technological advancement and the threat of nuclear destruction.

There are many who might wish that two suns surrounded us with perpetual light so that we might never have peered into the darkness for the advances that mark the progress of civilization have as often as not been one person's gain at the expense of another. And the possibility of nuclear destruction weighs heavily in the balance.

Every tribe of man has seen fit to slaughter, enslave, or sacrifice another, no matter how noble the savage. If the ways of the past were to replace the ways of the present, the world's population would be one-tenth of what it is today; and it is unlikely that you and I would be around to do the counting. People have no doubt found more comfort in the organic intimacy of the gods of the wood than from within sanitized modernity. But few of us would seriously choose to return to the darkness and ignorance that preceded the Age of Enlightenment.

There is room for optimism, however. Some of us believe that it is possible to have the best of both worlds. Modern civilization cannot be looked upon as good or bad—not yet. It is, after all, a great experiment, still unfinished. We have no choice but to follow it through, to strive to fuse today's wealth of knowledge and material goods to yesterday's organic intimacy. We must continue to look to the stars

THE CONSTELLATIONS

People often have a lot of trouble finding the constellations. It's so difficult to take connect-the-dot pictures of stars on paper and see them in the sky that most can never find anything more than Orion the Hunter and the Big Dipper at best. But the most spectacular constellations really do look exactly like their names: Leo the lion, Scorpio the scorpion, Cygnus the swan, Pegasus the winged horse. And better still, most of them can be found by using simple, ever-present skymarks.

What follows are several tales of mythical Greece and Rome, all filled with characters that are easy to find in the sky. Just as they occur together in legend, so do they appear together in the sky; once you've found one, the rest come easily. But remember, people see things differently. Don't feel that you have to see the same image that the artist does. You may even find it easier to remember a constellation in a shape wholly unlike the name given it.

How did these creatures get up there? Are they dangerous? Are they being fed properly? Don't worry, there are perfectly logical explanations for these and other vital questions . . . They may be found by going back a few thousand years ago, to ancient Greece.

THE HUNTERS AND THE HUNTED

Winter nights are the best for stargazing. This is the season of the most brilliant constellations, the clearest air, the longest nights, and the most expansive views. But it's cold! So make yourself comfortable. Take along blankets and hot chocolate. Don't feel guilty about temporarily retreating to the house or car. If you must turn on a light, mask it in red using tape, tissue paper, or marking pen, so that you can return to stargazing with your night vision intact. Swinging your arms in circles gets the heart pumping and whips blood into the fingertips, an excellent method for retaining the warmth in your hands.

The best time for viewing the Hunters and the Hunted is from the middle of January through April. The Big and Little Bears, since they are among the never-setting circumpolar stars, can be seen year-round.

The king of the gods, Jupiter, was known to be a bit of a lady's man among the pretty mortal maidens he chanced to see on his trips to Earth. This didn't sit too well with Juno, who happened to be his queen. One day, as Juno was looking through the clouds below Mount Olympus, she spied one of Jupiter's mistresses in the forest. It was Callisto, taking her morning constitutional.

Zap! Juno changed her into a bear. When Callisto cried for help, she roared like a bear. When she tried to go home, she realized that she smelled like a bear, too. The poor woman wandered the woods on her hands and knees, until one day a young man searching for the missing Callisto spotted her. "Could this bear have eaten Callisto?" he thought. "Callisto, my *mother*?"

Callisto rose up and cried out to her son. The boy drew his bow, aiming at her heart. But wait! Jupiter swept down from above, and in one mighty stroke, he transformed the young man into a bear. Then Jupiter grabbed both bears by their tails and cast them among the stars. This is why the bear constellations have

Jupiter hurls Callisto into the heavens.

such unnaturally long tails. (Like I said, there's a logical explanation for all this stuff.)

Well, Juno was in a snit! She stormed off to her parents, Tethys and Oceanus, grand poo-bahs of the seas. She demanded that they never permit mother and son to drink of their waters. Not even a little bit, not even to wash off that bear stink. They agreed, and so it is that the Big and Little Bears are confined to the very center of the sky, around which everything else revolves, and are never permitted to dip below the horizon and into the seas.

These and all the other stars that are forever above the horizon are known as the circumpolar stars. The Big Dipper forms the tail and hindquarters of the Big Bear. It is easier to find if you first locate the North Star by following the red arrow on your compass to north and going up roughly four fist widths.

Now look for the Big Dipper circling around it. It's about three fist widths away from the North Star and is two and a half fist widths across. At the time of year that the Hunters and the Hunted are best seen, the Big Dipper either will appear to be pouring from the left to right or will be about level. The stars that make up the rest of the Big Bear are not as bright as those in its dipper part, but they clearly outline a large constellation in the shape of a bear. (page 35)

The Little Bear is one and the same as the Little Dipper, the North Star being the tip of its tail (or handle). It takes some imagination to see a young cub here, but there's no problem identifying it as the Little Dipper. It's important that you are able to find the Big and Little Bears (Dippers), because they are used to find many of the other constellations.

Now take the two stars that complete the far edge of the bowl in the Big Dipper, the ones that point directly to the North Star. Shoot them four fist widths in the opposite direction. This will take you straight to the heart of Leo the lion.

Parallel to and facing the same direction as the Big Bear, Leo is angled to pounce on some unsuspecting varmint. He's about two and a half fists across and just over a fist from the bright star Regulus in his chest to the top of his head, an area that resembles

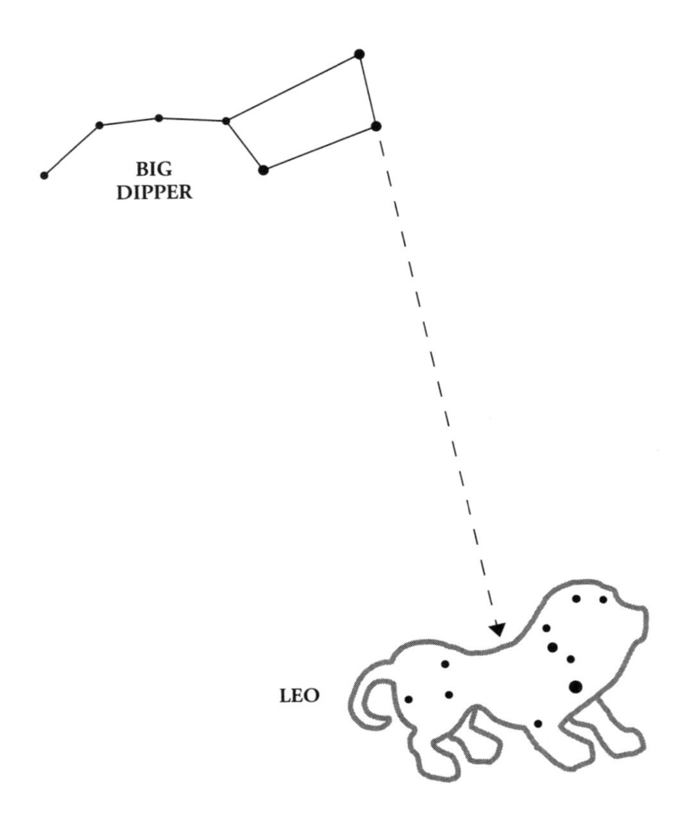

a backward question mark. Leo is not only one of the twelve signs of the zodiac, but he's also the lion whose destruction was the first of the twelve labors of Hercules.

Leo was born of the moon and fell to the Earth as a shooting star. Scarred from his dysfunctional childhood, he roamed the Nemean forest, attacking everything in sight, whether it needed it or not. No hunter could kill him, for not only was he as large as a house, but he was very thick-skinned; spears, arrows, and sarcasm bounced right off. Hercules had to sneak up on him and club him with a huge tree, then strangle him to death with his bare hands. (Hercules had very large hands.) Then he skinned Leo and hung his hide in the heavens as a trophy. Hard to believe? Look in the sky and, yep, there he is.

Now look to the south. See the three bright stars in a row, the width of two outstretched fingers? These form the belt of Orion the Hunter, the most brilliant of all constellations. Now, imagine where a man's shoulder and legs would be (about two fists from shoulder to thigh), and you'll find a brighter star at each point. You may notice that the star in his left shoulder, Betelgeuse (bee'-tel-jooz), is red in color. You can even see a faint sword of stars swinging from his belt.

Orion (or-eye'-on) was a great hunter who had cleared the wild beasts from the kingdom of Chios, a stone's throw from Nemea. Diana, goddess of the hunt, thought he had nice bone structure and fell in love with him. Together they roamed the woods, hunting and learning about the birds and the beasts.

Apollo, brother of Diana, was afraid that she would break her vow never to marry and schemed to split them up. One day, as brother and sister sat up on Mount Olympus, Apollo saw Orion fishing in the sea, wading deeper and deeper into the water until only his head showed. Apollo challenged his sister to see if she could hit the black spot on the waters. Not only was Diana gullible, she was a good shot, too, and she pierced Orion's head with her arrow. She was horrified when she saw his body wash ashore. The grief-stricken Diana placed the body of her beloved (minus her arrow) in the sky for all to see.

His two hunting dogs—the Big and Little Dogs—follow him. It is often difficult to see anything but one bright star in each, but those are hard to miss. Sirius (seer'-ee-us), the Dog Star, is the brightest star in the sky and is just over two fists to Orion's left, on a straight line drawn through his belt. Procyon (pro'-see-on) is the same distance on a line drawn through Orion's shoulders (drop down a bit toward Sirius). The three stars Betelgeuse, Procyon, and Sirius form what is known as the Winter Triangle.

Orion and his dogs are chasing after the Seven Sisters, or Pleiades (plee'-uh-deez), a beautiful but faint cluster of stars also on the line through Orion's belt, but to the right by four and a half fists. (page 46) They are the daughters of Atlas, and they shrieked for help as damsels in distress are wont to do. Atlas was busy shouldering the globe at the moment, so he called on Jupiter. Jupiter also had an eye for the sisters, but he knew he couldn't afford any hanky-panky with the daughters of the man who had the whole world in his hands. *Zap!* Jupiter changed them into seven pigeons, and they fluttered up to the stars.

Jupiter was serious about keeping Atlas at his work, so he set Taurus (tor'-us) the bull between Orion and the Seven Sisters. This V-shaped constellation appears as only the head and horns of a bull. The bull's eye can be found midway along the line running from Orion's belt to the Seven Sisters—you can see its angry red glare. (The tips of the horns are a fist and a half above, as though Taurus is crashing down on Orion.) Leo and Taurus are both constellations of the zodiac.

Taurus is none other than Jupiter himself, up to his old tricks. He fell in love (again) with a beautiful mortal girl, Europa. Afraid

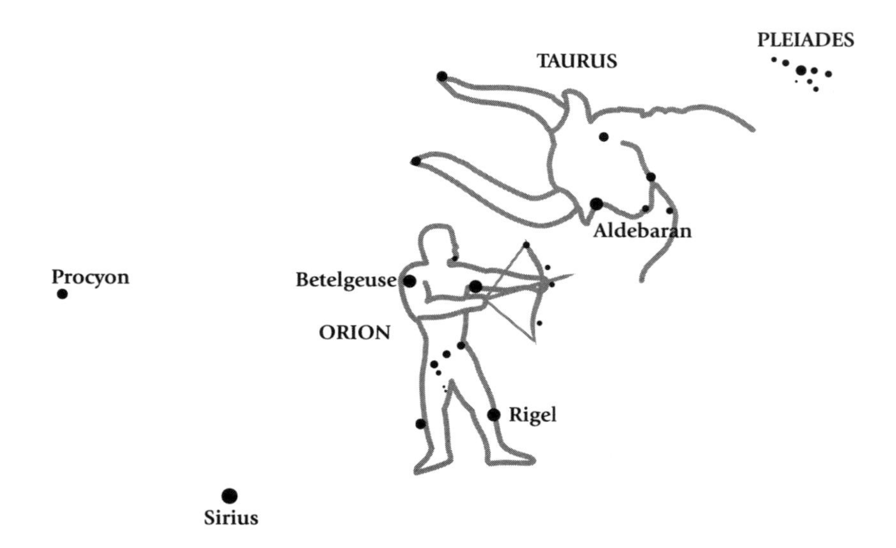

The key to the winter stars is finding the three bright stars in Orion's belt, in the southern sky. They're in a straight row the width of one outstretched finger.

that Juno would see them and change her into a wombat, he disguised himself as a bull and roamed among Europa's herd of cattle. Europa was taken by the handsome, friendly beast. She climbed onto his back and boy, did she get a ride! He trotted off with her into the sea, where he swam the waters to Crete. This is why the constellation shows only the head of Taurus. Once in Crete, Jupiter resumed his immortal form to have an affair with Europa, who was astonished, to say the least.

THE SUMMER TRIANGLE

There is another prominent triangle in the sky, the Summer Triangle. You might guess that that's the best time to look for it, but you're wrong. (Hah!) The Summer Triangle is best seen in the fall, directly overhead, although it's visible from midsummer to early winter. There are a few simple clues to finding it. One is to simply look for a large triangle, roughly three fist widths on each side,

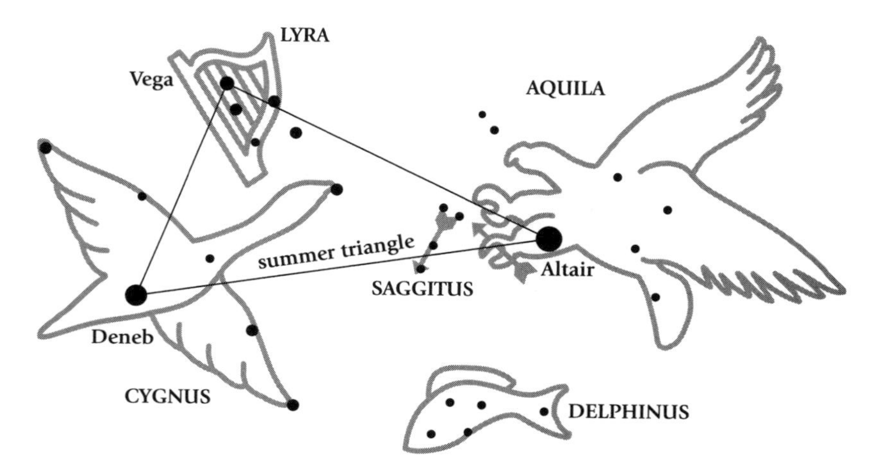

The Summer Triangle. In the fall, three bright stars can be seen directly overhead: Deneb, Altair, and Vega, roughly three fist widths apart from each other.

that is formed by the three brightest stars directly overhead and a little to the west. If you're in the country, you can see the Milky Way galaxy streaming through the triangle as well.

These three stars are the brightest in their constellations, two of which are the beautiful outlines of birds. The third is a lyre, a type of harp. Accompanying this triangle are two other constellations, an arrow and a dolphin.

Aquila (ak'-will-uh) is the eagle appointed to carry Jupiter's thunderbolts. Saggitus (sa'-ji-tus), the arrow, is one of them, and if you look into the sky, Aquila seems to have dropped it. Don't believe me? Then take a look between Aquila and Cygnus. If that's not an arrow, then I don't know what is. Maybe Aquila was just having a bad day, because in the following story, he's up to the job.

Hebe was cupbearer of the gods. She continually replenished their goblets with the Olympian ambrosia (not entirely devoid of alcoholic content). Once, while carrying three pitchers in each hand, Cupid launched an arrow at her rear. Hebe shrieked and slopped ambrosia all over Jupiter, who had been chatting with a young nymph.

Jupiter jumped up, spluttering and wringing out his beard, ready to give her the boot. But it mattered not. Hebe's eyes had fallen on the nearby Hercules, and she floated to him like an antelope slowed in midleap. Jupiter struggled to recover his composure in front of the nymph. He summoned his swiftest servant, Aquila. "To Earth—immediately! Find me a sturdy lad who can hold his liquor—and mine, too."

Aquila dispatched with haste and plucked pretty-boy Ganymede from the midst of his dive competition. Aquila swooped the flabbergasted lad back to the Olympus Bar & Grill just in time to refill Jupiter's goblet. Since the grateful Lord of All Gods rewarded Aquila with a place among the stars, we can only imagine how things went with his nymph "friend." And as any astronomer will tell you, Ganymede continues to orbit around Jupiter.

Altair (al-tare') is the bright star in Aquila's head, and of the stars in the Summer Triangle, it is the one farthest from the Big Dipper. Aquila is two fists long from head to tail and just over two fists from wingtip to wingtip.

Orpheus was given Lyra (lie'-ruh), the lyre, by Apollo, and he played so well that the birds of the air, creatures of the sea, and beasts of the forest would turn to him wagging their tails and tapping their feet in unison. Ferocious animals, on hearing his sweet harp, would gag in midroar; the lion would cease ripping on the lamb, who would die a lingering death.

Fair Eurydice fell in love with Orpheus and his lyre, and they wanted to make beautiful babies together. But alas! A snake, deaf to the charms of the music, bit bunny-wonderful Eurydice on the heel, and she died. An untimely death, as ever there was. Orpheus's heart was rent (that means split in two, not that it's the end of the month). He played so sadly on his harp that the skies cried buckets of rain, the earth quaked great sobs, and arrows shied from their intended targets. Consequently, many people died from floods, collapsing structures, and stray arrows.

Orpheus beseeched stony Hades in his Hell to release beloved Eurydice, plucking the charms of his magic strings. Hades himself wept so hard that he quenched his ravenous fires. But he was grateful for the new additions, so he consented to Orpheus's request, on one condition: As Orpheus led his schloopy, Eurydice from the Underworld, he might not turn to face her until they had emerged into the light. (Uh-oh . . .)

Orpheus rejoiced! As soon as he felt the hand of his honey-sweet clasped within his own, he led her upward. He strummed a happy little ditty on his lyre, and the whole world sang with him. But forsooth and forswear! He turned to his cupcake to join in the refrain. She moaned, her hand shrank within his, and she faded away, cooing Orpheus's name.

Poor Orpheus! And once again began the most plaintive, heartsick strumming that ever was heard. But people were tiring of his sappy melodies. Hades glared unsympathetically out of his

Underworld steambath. People were hoping that maybe Orpheus would have an "accident." But the stones and spears launched his way fell at his feet, for still his music worked its sad magic. But weep not, kind reader. Ours is not a tale of woe—read on!

Hades schemed mightily, rubbing his chin. Then he snapped his fingers—the Furies! Winged hellions with the fangs and claws of large housecats, they could do the job. They surrounded Orpheus and his lyre. Jamming their fingers into their ears, they screamed, "We can't hear yoo-o-o!" and tore him to pieces with their taloned feet.

So our story ends, with Orpheus finally rejoining Eurydice in the depths of Hell, where they died together happily ever after. The gods placed Lyra in the heavens, that it might never be played again.

Vega (vay'-guh) is the brightest star in Lyra; actually, it's a double star. It's the closest of the three Summer Triangle stars to the line coming from the bowl of the Big Dipper or, if you're facing north, the leftmost of the two triangle stars that are closest to north. The rest of the constellation is comparatively dim, forming a pretty diamond of four stars under one fist across, pointing toward Altair in Aquila.

We can't tell the story of the third constellation in the Summer Triangle, Cygnus (sig'-nus) the swan, without telling a story of the zodiac as well. But that's just as well, because this chapter includes five of the zodiac's constellations.

Cygnus was buddies with Phaëton, son of Apollo. Like many uneducated people today, their schoolmates didn't believe that Apollo actually drove the chariot of the sun across the sky, let alone that Phaëton was his son. Apollo's job kept him on the road so much that Phaëton hardly ever saw his father, and the boy grew tired of being teased in school. They called him horrible, mean things like Sparky and Not-So-Bright, which would send his loyal friend Cygnus raging after them.

One day Phaëton decided he had had enough and marched up to Apollo's palace. "I want to drive the chariot of the sun so all my friends can see me," he shouted, stamping his foot. "Now!"

Apollo's sunny disposition clouded. "No, not that! *Anything* but that!" This was worse than putting a teenager on your insurance policy. But Apollo felt guilty for not being there for his son and wanted a day off himself.

The gates were flung open and the snorting chargers stepped forth, the sun rolling behind them. "Not too near the Earth or too high," Apollo hollered at Phaëton as he climbed into the chariot, but too late. They were off!

The cart surged with the lightened load, and as they overtook Virgo, her tresses were singed. Phaëton could not steer the raging steeds along the zodiac, and upward they shot toward the Big and Little Bears, burning them into black bears. Scorpio whirled, lashing his barb at the chariot and so startling the horses that they wheeled toward Earth. Atlas shuddered from the heat, spilling lava out of the Earth. The sun, whipped around behind Phaëton and his chargers, incinerated entire cities and left rivers sizzling in their banks. Africa turned to dry desert and her people were singed dark, while the peoples of the Mediterranean turned white with fear.

The fire chariot careened back and forth, and the Gemini Twins whooped and danced to save their feet. Taurus bent low, snorted, and charged, forcing the horses toward Pisces the fish, baking them into a blackened roughy. Draco breathed fire from his nostrils, but missed the wild caravan and evaporated the water that Aquarius was ready to dump on them. The world was on fire, and the sky was falling!

Jupiter raised a wicked bolt overhead and flung it across the blazing heaven. The chariot exploded, the horses raced free from their burden, and Phaëton fell headlong into the River Eridanus.

Cygnus had watched it all from below. He dove in to save his beloved friend, but the waters were too clouded from the falling embers of the zodiac to find him. Again and again he plunged below, until long past the time even a Japanese pearl diver could have held her breath. The gods took pity on Cygnus and changed him into a swan. Now he searches from the sky, diving into whatever waters lie below, eternally loyal to his friend Phaëton.

The end of Phaëton's wild ride.

Deneb (Den'-eb), the third corner of the Summer Triangle, is the bright star in the tail of Cygnus. Of the two triangle stars closest to north, this is the one on the right when you're facing north. Cygnus has to be one of the most elegant constellations, with stars detailing every contour of his body. He appears to be flying in the opposite direction from Aquila. He is two and a half fists wide and three fists long.

Even the ancient Greeks considered the dolphin to be quick-witted and light of spirit. Thus, even though the sky has a region filled with sea creatures, called "the Ocean," they chose to place him among the constellations that flew through the air.

Neptune, god of the sea, began courting the water goddess Amphitrite. But she was coy to his advances, winking shyly and hiding behind a grouper. So Neptune sent his pet, Delphinus (dell-fee'-nus) the dolphin, to charm her. The playful beast would leap about for her, balance a ball on his nose, smack her on the lips, even pull her on skis. Then suddenly, he left.

Amphitrite missed Delphinus sorely, knowing not whither he had gone. Imagine her surprise when one day he returned, plunging through the waters. Only now, Neptune rode astride his back, one hand holding the reins, the other stretched high overhead, the god smiling through his seaweed beard. So taken was Amphitrite by this routine that she immediately fell in love with Neptune. They married and lived happily ever after.

Delphinus is fainter than the two birds flying over it, but its stars are all about the same brightness, and they really look like a dolphin. If you can imagine a fourth corner to the Summer Triangle that completes a rectangle, across from Vega, you'll find Delphinus. It's only three fingers across.

THE CLASH OF THE TITANS

Spread across the sky are six constellations that form the story of the great hero Perseus (pur'-see-us); they may be seen from mid-November until the end of January. No other myth attains such astronomical proportions, for all its cast of characters is preserved in the sky. This makes both the legend of Perseus and the stars in his honor easy to remember. Not all of these are easy to find or resemble their namesakes; we'll concentrate on those that do.

Perseus was the son of Jupiter. At birth, he was shut up into a chest with his mother and cast into the sea, the result of another creepy Greek oracle that prophesied that the child would kill the father. The fisherman who found it gave it to his king, who raised the child with hopes that one day the sturdy youth could slay the Gorgon Medusa.

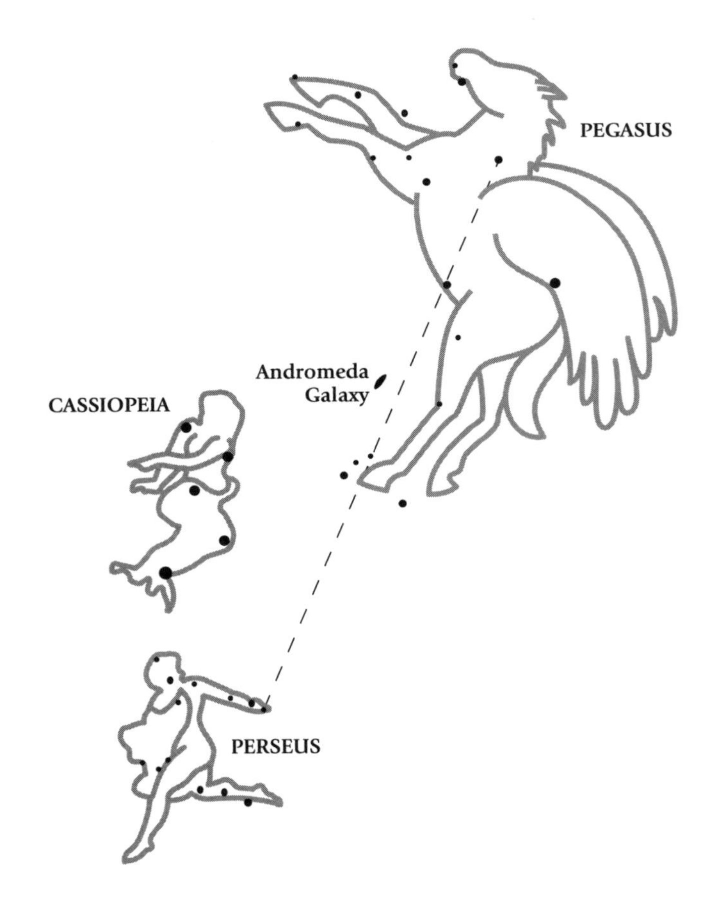

Once Medusa had been a comely maiden (a cute virgin) who dared to compare her beauty with that of Minerva, goddess of wisdom. Once pretty, *Zap!* now pretty ugly. Minerva transformed her into a cave-dwelling she-monster, so hideous that whoever saw her turned to stone.

The big day came for Perseus to slay Medusa. Mercury, the god of travelers, loaned Perseus his winged sandals. Minerva (in a clear conflict of interests) loaned him her highly polished shield. Now Perseus could see Medusa's reflection and do battle without turning into classical sculpture. Quick like a bunny Perseus fluttered to his rendezvous with destiny.

Statues of men surrounded Medusa's lair, warning Perseus that she was near. Brave, young Perseus boldly flitted into action, sandal wings flapping, careful to look at Medusa only in Minerva's shield. He swooped in and lopped off her head, which he dropped into a bag. The blood that shot out from the gaping, revolting stump of her hacked-off head drizzled to the ground. From this sprang the winged horse Pegasus (peg'-uh-sus).

Now in the land of Ethiopia lived King Cepheus (sée-fee-us) and his queen, Cassiopeia (kass-ee-oh-pee'-uh). Cassiopeia had made no secret of her beauty (bad move) and let it be known far and wide that she was so foxy that even the water nymphs couldn't compare. (Nymphs are like the Sea-monkeys you buy in the back of comic books, only bigger, and they're for real.)

The nymphs sent Cetus the sea monster to devour the coast unless King Cepheus would appease the nasty nymphs by sacrificing his daughter Andromeda (an-drau'-meh-duh). Well, if he had to, he had to. They marched Andromeda down to the coast, chaining her to a cliff. Cepheus and Cassiopeia were torn between returning home to mourn or staying to watch their daughter eaten alive. They stayed.

The tides surged in, the dreaded sign of the monster's approach. Cetus, the virgin-eating monster whale, came closer — and closer—and CLOSER!

But look, in the sky—it's Perseus on his way home with Medusa's head! The valiant youth spotted the terrified and not at all bad-looking Andromeda chained below in her spray-drenched bodice. Noble Perseus knew he must stop. A good thing, too, because the gigantic sea monster was at that very moment about to eat her. Perseus drew his sword and zoomed at the monster, stabbing this way and that. It was a *big* monster, and he stabbed it many times, each time narrowly escaping death or worse. But wait! Perseus remembered something: All sea monsters are particularly vulnerable in the upper neck, where the first vertebra enters the cranium (Hero School 101). He rejoined battle, dealt the death blow, and freed Andromeda. The proud foursome returned to the castle, leaving Cetus to rot on the coast.

Perseus and Andromeda married, of course, but Perseus let Medusa's head out of the bag during the ceremony and froze half the wedding party. That's another story that spoils an otherwise happy ending, and we're not going to tell it.

If you continue the line of the Big Dipper's pointer stars through the North Star, about five fists and a little to the right, you will arrive at Cassiopeia (six fists from the pointer stars) and then Pegasus. Cassiopeia appears as a W of stars one and a half fists wide, opening toward the Big and Little Dippers (Bears). It's easiest to picture this W as a throne with Cassiopeia seated on it.

Extending the pointer star line another three fists beneath the throne, you will arrive at the Great Square of Pegasus, roughly one fist on each side. This forms his body, and it is possible to see hindlegs, forelegs, neck, and head, all in proportion. The line you have extended strikes him in the belly, and he appears to be galloping to the left if you're seeing him right side up (facing the south).

Andromeda is one and the same with the hindlegs of Pegasus, and this is the location of the only other galaxy besides our own that we can see with the naked eye, the Andromeda galaxy (see the end of chapter 7). If you draw a line from the shoulders of

Pegasus through his rear thigh (from one corner of the square to another) and continue it just under five fist widths, you will arrive at the bottom of Perseus. This is a diffuse stream of stars with no clear shape. Though neither Perseus nor Andromeda look much like their names, they are easy to find. Both Cassiopeia and Perseus lie in the trail of the Milky Way.

THE HERDSMAN, THE NORTHERN CROWN, AND THE ZODIAC

The constellations of summer are not especially bright, but the warm weather encourages one to lie down on a blanket and comfortably pick them out. Though these constellations do resemble their namesakes and are easy to find, the myths surrounding them aren't great epics like the earlier ones. These summer constellations are best seen from mid-June through late September.

Boötes (bow-owe'-teez) is the herdsman, shepherding the Big and Little Bears along their appointed path. The gods realized how valuable these stars were to navigation, and even Jupiter gave up his anger at Juno for never allowing them to dip into the seas and set Boötes to his task.

Boötes may easily be found by extending the arch of the Big Dipper's handle three fists until you find the bright star Arcturus (ark-tur'-us). This is where his legs join his body, which forms an elongated diamond (his torso) two and a half fists long, stretching toward the north.

Coming off Boötes' right shoulder is a constellation of beautiful simplicity. This is the Corona Borealis (kor-oh'-nuh boar-ee-al'-is), or northern crown. The god of wine and revelry, Bacchus, found the princess Ariadne abandoned on a deserted island and gave her this crown as a wedding present. It is possible to see the crown as Boötes' arm, bent in a graceful semicircle.

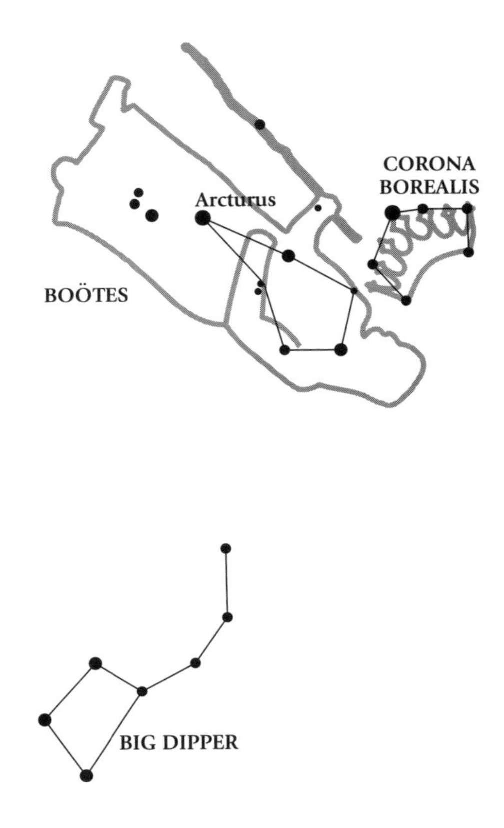

There is a constellation that bears no resemblance to its name whatsoever, but looks exactly like a teapot. Its stars are those of Sagittarius (sa-ji-tare′-ee-us) the centaur, half man, half horse. He was put in the sky to protect Orion from the sting of Scorpio. But even as kitchenware, Sagittarius defends Orion still, for the teapot pours scalding water on the scorpion (the Milky Way appears as steam coming from the spout). When looking at this point, you are looking into the center of our galaxy.

The centaurs were accepted as equals with men, so highly did the Greeks esteem the horse. Humans and centaurs shared alike the Earth's bounty until one wretched day. The centaurs were invited to a wedding and arrived in all their leather finery. The ambrosia flowed freely, and there was much mirth and merri-

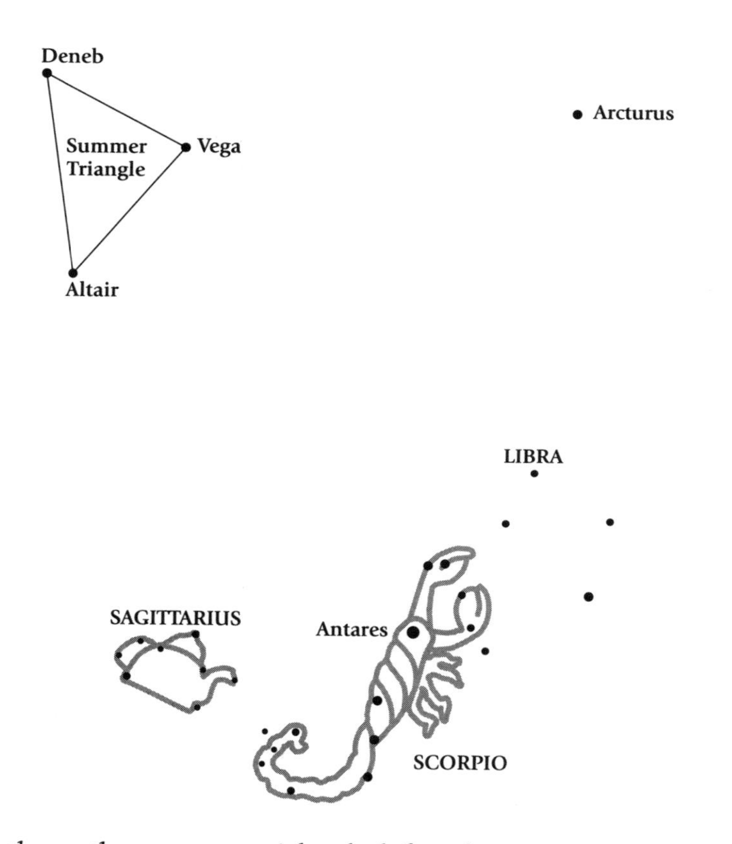

In the south on summer nights, look for a huge scorpion standing on his tail, three fist widths in height.

ment. But then the centaurs tried to kidnap the bride and her maids and have their way with them, no doubt the result of some confusion over biology. This led to a battle with their human hosts in which several centaurs were killed.

Not all the centaurs were so ill-behaved, however. Chiron was a true Renaissance horseman, learned in the arts, prophecy, and medicine. He became the mentor of Aesculapius, the father of modern medicine, and it is Chiron we must think of when we see the teapot in the sky.

Scorpio the scorpion looks just like his name. As Orion, his foe, rules the southern skies of winter, so Scorpio rules the summer. Between his two claws is another constellation of the zodiac, Libra (lee'-bruh), visible only as a box of four fainter stars. All three of these zodiacal constellations appear in a row, Sagittarius-Scorpio-Libra, just above the southern horizon. It may not be possible for people in more northerly latitudes to see them. Sagittarius is one and a half fists across, and Scorpio sprawls three fist widths. Libra is only two to three fingers on a side, barely visible as a box within the arc of Scorpio's claws.

THE FUTURE IN THE SKY

We all recognize our dependency on the sun for heat, light, and the seasons. But the sun has other vitally important effects, as do the moon, planets, and denizens of deep space so far away it's amazing the power they have over our lives. The universe is the environment the Earth we live on lives in, as surely as a deer lives in the woods or a fish lives in the sea. In every field, scientists are finding that if they fail to account for the influence of distant pinpoints of light, their models fail to truly represent our world.

This is why scientists so carefully studied Comet Shoemaker-Levy in 1994. It crashed into Jupiter, leaving an impact site far larger than the size of the Earth. If such a tiny object from so far away had struck our planet, it would have made the existence of most life impossible.

Even more disturbing is knowing that Earth's gravitational field actually *attracts* comets and other debris. In 1908, what was probably a comet broke up when it entered the atmosphere over Tunguska, Siberia. It flattened trees within an area 40 miles across. Fortunately, it occurred so remotely that nobody was killed, and it wasn't formally investigated until nineteen years later.

But don't worry about the Earth's being destroyed by a comet. It's a million times more likely to be destroyed by a meteorite, if only because there are a lot more of them. Though a comet's

As surely as a fish swims in the sea so, too, is our Earth a part of the universe around it.

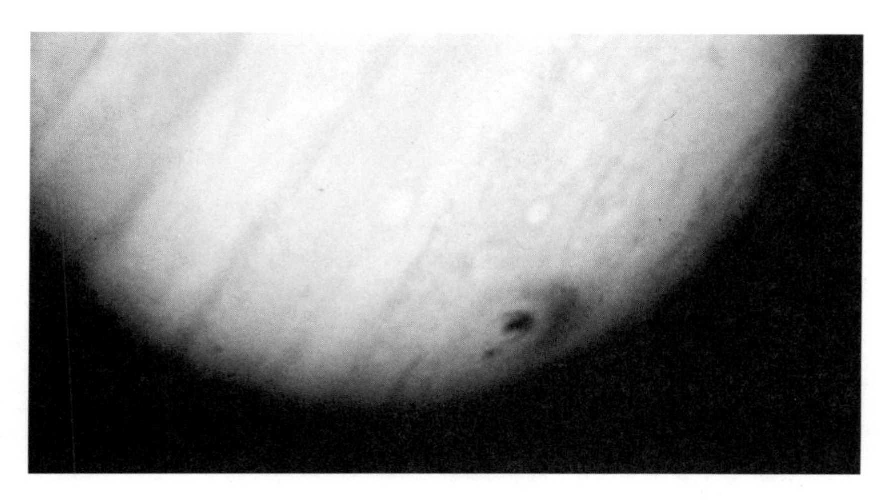

Comet Shoemaker-Levy's impact site is at the lower right of this photograph of Jupiter. Earth would fit inside of it.

5-mile nucleus is much bigger than the average pea-size meteor that streaks through the sky and burns to ashes, they're outnumbered by shooting stars a trillion to one.

After all, a meteoroid is any space debris that enters Earth's atmosphere: asteroids, a falling communications satellite, cometary fragments. A very real concern among scientists (and everyone else on Earth) is that a huge asteroid will be pulled out of the belt between Mars and Jupiter and speed toward Earth, pulled in by our gravity.

Skeptics need only to look at the moon. It's covered with thousands of huge craters. The Earth, if only because it's bigger, has absorbed many more hits. So why is the moon teeming with craters and the Earth has only a few? Don't think it's because we're lucky or the moon is especially unlucky. It's because we have water, an atmosphere, and tectonic activity, all of which act to erode the countless craters all over Earth.

Ironically, the water in our oceans may have come from the ice of comets, and it could be the very frequency of devastating water-bearing comets that made life on Earth possible. Stranger still,

Heavily cratered and ever showing the same face, the moon has tugged at our oceans and imaginations for a thousand generations.

there's speculation that it's because of two huge meteor strikes that the dinosaurs became extinct and the Earth has the people-friendly climate it does now.

Hundreds of tons of tiny meteorites pound the Earth every day, and some familiar places were created by gigantic meteorite strikes: Hudson Bay in Canada, the Gulf of Mexico, part of the Indian Ocean, and the Sierra Madre in Texas.

While death from the sky deserves our consideration, we are somewhat protected by the planet Jupiter. Jupiter is so much more massive than Earth that it draws to itself many of the asteroids, comets, and space debris that might come our way. Until Comet Shoemaker-Levy crash-landed on Jupiter, it had been regularly orbiting past Earth.

People have held a wide range of opinions as to the evil or good effects from objects in the sky. Some british sailors used to believe that if they slept on deck, the rays of the moon could poison them. The Amish, among others, believed that haircuts are best had when the moon is waxing (growing from crescent to full) and that crops should be harvested only when the moon is on the wane. Fish markets advertise that shellfish are tastier and larger when harvested under a full moon.

Sailors may sleep in peace, and farmers may reap when they will. If these things occur, it's in extraordinary times and places. But the effects of the moon on marine life are well documented. Herring runs and other fish migrations vary with the moon, as do the reproductive cycles of many marine mammals. Harvested shrimp are plumper, and jubilees, in which thousands of fish float to the surface, doing nothing to avoid capture, occur in phase with the moon. The fact that these effects are most prominent among creatures of the tidal zone, the narrow strip between land and sea, gives some clue as to why the moon has this mysterious effect: the tides.

Gravity, one of the forces underlying everything we know about the universe, again comes into play. Just as the Earth holds the moon captive at its side, so, too, does the moon tug at the Earth. But this force is less powerful and exerted unequally. The Earth's crust is distorted by only 8 inches as it spins beneath this force, although the fluid oceans may flow tens of feet higher. Every ocean visitor can observe this effect in a few hours' time.

It follows then that sea creatures would adapt their living patterns to this powerful cycle, which defines how much sunlight they receive, what their habitat is, the temperature, the abundance of food, and so on. It is very similar to our rising and our daily activities being governed by the sun.

We've seen earlier how the sun contributes to the tidal effect. When the sun and moon are in alignment with the Earth, the tides can be half again as strong. These extremes churn the oceans and are strongly suspected to cool global temperatures. Earthquakes are similarly linked to this pull of the sun and moon. Our closest neighbors don't merely pull on the surface of the Earth, but on its entire body.

This distorting force has pulled the Earth out of round, a fact that made it very difficult to navigate and map the Earth in the eighteenth century. It's not a continual tug on a straight line drawn through the Earth. It's a wave that girdles the Earth and rolls around it every day. It should come as no surprise, then, that the moon and sun are contributing sources of earthquakes.

Are there more subtle influences from the moon and sun? "Lunacy" comes from "lunar" or "of the moon." People have quite logically believed that, just as they observe the moon swaying huge Earth, surely it must influence lesser people and animals. Logical, but incorrect. The frequency of suicide, wolf packs gathering, and people chopping up baby-sitters does not increase during the full moon. This is lore and romance, although there are people who have a special sensitivity; women who spend a lot of time outdoors may report menstruating in lunar synchronization.

The moon certainly affects the behavior of animals, but most often because they *see* it, not because they are in some special sympathy with its powers. Dogs howling, birds migrating at night, plants opening, and amphibians reproducing are not so very different from our own tendency to work when there is light and to sleep when there is darkness.

The sun has other effects still that could dramatically change our lives. To look into the future, let's go back to the past, starting in the early seventeenth century.

In 1610, Galileo reported that, using his telescope, he could see black spots on the sun. He drew detailed charts showing the weekly progression of these spots. This violated the Church doctrine that everything was created by God and was therefore perfect.

These tiny, intangible sunspots, along with Galileo's other discoveries, were the greatest challenge to the authority of the Church up to that time. The Holy Fathers could excommunicate Luther and wage war on the Muslims, but they could not intelligently deny what everybody could plainly see through the telescope: the sun had spots.

The seventeenth century was not half over before the entire world experienced a great drop in temperature that lasted until the early 1800s. The bitter cold reached so far south that thousands

The advance of science was slowed by leaders who refused to let others speak or hear of what they themselves could plainly see—the sun had spots.

froze to death, ships were trapped in ice, and many more starved. And those who followed the Church Fathers pointed to the sun and proclaimed, "There are no sunspots, and there never were," and claimed that Galileo's heresy had brought the wrath of God upon them.

And they were right: there were no sunspots, and even the unfaithful wondered if there ever had been. What had happened?

Sunspots are intense magnetic areas that occur in regular pairs on either side of the sun's equator. These are relatively cool regions, but they are precisely where the huge and violent solar flares erupt. These bursts of superheated particles shoot toward the Earth and add energy to our atmosphere.

The mysterious absence of sunspots was responsible for the great cooling, also known as the Maunder Minimum. It is uncertain if this occurred randomly or is part of a regular cycle. Either way, it is likely that it will happen again. This may be one reason

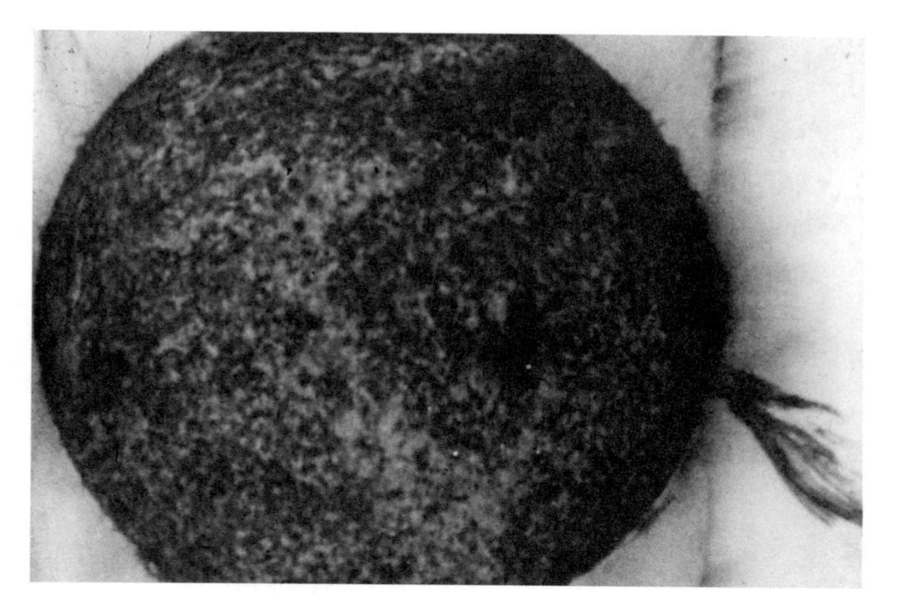

The brilliant jet to the right is a solar flare, thousands of times the size of the Earth.

the Vikings abandoned their three-hundred-year-old settlements in Labrador, delaying the European colonization of America.

These solar flares not only make things warm and toasty, but their heating of our atmosphere increases the drag on satellites and helped pull *Skylab* out of orbit in 1979. If satellites are not carefully monitored and reprogrammed during periods of intense solar flares, they can fall out of the sky.

This bombardment of charged particles is what causes the magnificent auroras, and it is also a common source of communications interference. This is serious enough that an organization called the Solar Terrestrial Dispatch constantly collects and posts warnings of auroras and other electromagnetic interference. If your cellular phone goes out, this may be why.

Solar flares have been implicated in contributing to flooding worldwide, and they were also directly responsible for a major disaster close to home. In March 1989, a huge solar storm knocked out the entire power grid of Quebec. This left nearly a third of Canadians (6 million people) without power in the late winter. Power was lost or reduced elsewhere in the U.S. Northeast and Southwest. The auroras, needless to say, were beautiful.

The poor dinosaurs have been killed off by so many different theories that it would be positively unsporting not to suggest another one: a supernova explosion. Beyond the reaches of the moon and sunspots, in a galaxy far, far away, a star was dying. It could no longer churn out enough nuclear energy to overcome the gravitational tendency to collapse under its own weight. The huge implosion that occurred is thought by some to have created an X-ray bombardment of the Earth so furious that it wiped out the dinosaurs.

What's disturbing about this is that even if it isn't what killed the dinosaurs, it easily could have, and it could happen anytime. The only reassurance we have is that as far away as the stars are, anything that takes place in outer space will probably take a long time to get to us. Then again, it could already be on its way.

What might happen when the explosive force of the Big Bang gives out? Will gravity take over again and crush the universe into a basketball? No one knows. Will the sun expand into a red giant and engulf the Earth in a nuclear inferno? Yes, but not anytime this week. What we see in the sky has had and will continue to have tremendous influence on the Earth, and we know that there's a lot we don't know.

Modern man looks up and see his future in the stars. Ancient man looked up and saw his future in the planets. Great observatories were built around the world in order to inform the king and his subjects of what the heavens declared. These architectural wonders were used as machines to calculate eclipses, highs and lows of the sun and moon, and the positions of key planets and stars. The priests overseeing these holy observatories knew that these motions were predictable, that there was not an unseen, sentient force mysteriously arranging the stars and planets according to divine whim or human virtue.

Yet the general population felt completely vindicated when, after having spent centuries building mammoth temples, the priests announced, "Jupiter will now retreat from its present course," and this came to pass. Of course, this was always followed by, "It will only return to its proper place when every village has brought a young man and woman to the temple."

This was the basis of astrology, the attempt to divine people's personalities and future by the stars. There are physical influences on people's personality that do have a distant basis in astronomy. People born in the winter months or in the higher (colder) latitudes are known to be moodier. But this has nothing to do with the planets or constellations, and no one has ever demonstrated anything that does. If there is one effect that the planets have on our lives, it's in their ability to persuade people to read their horoscope in the daily paper.

SIMPLE ANSWERS TO SIMPLE QUESTIONS

The problem with questions about astronomy is that the answers only inspire new questions, and very quickly these turn into metaphysical puzzlers. "Where does outer space end?" "When did time begin?" "Are we the only life in the universe?"

It is precisely this that accounts for man's timeless obsession with the stars. Of all his pursuits, stargazing is the one that can never be held, collected, written down, or mastered. Even if the stars had never held the key to mankind's greatest explorations and achievements, still they would fascinate us. Looking up from a beach on a starry night, even the youngest child understands that he peers into mystery. While the power of the oceans can be awesome, at least it is within the range of human understanding. The realm of the universe is simply incomprehensible.

As impressed as we are with the genius of Albert Einstein and the elocution of Carl Sagan, they may no more be considered masters of the topic than a mouse in a skyscraper is a master of architecture. If there is anything that sets them apart from the rest of us, it is only a deeper sense of awe in the face of how little they understand about the universe. It is this sensibility that has set them so earnestly working toward the Mystery. If you ever feel intimidated by the experts as they invoke dizzying principles and theories in their discourse on the heavens above, remember, you are only a slightly smaller mouse.

Do satellites ever crash into each other?

With more than six thousand satellites overhead, and many in similar orbits, someone's got to be looking out for them. For this reason, the flight path of every single satellite is painstakingly projected. With computers, the task of tracking the six thousand objects is not nearly as formidable as it seems.

Nonetheless, speed has a way of shrinking space, and many of these crafts are moving at 10,000 miles per hour. In 1997, there was a very real fear that a small piece of space debris would smash through the inhabited MIR spacecraft and destroy it. Luckily, many satellites can be guided from the Earth so that intercepts can be avoided.

Unfortunately, there's a lot of space debris up there: leftover rocket stages, abandoned receivers, discarded junk. Merely by virtue of orbiting the Earth, they also are satellites. Even though they can be tracked, that doesn't mean they can be guided, especially when one loses speed and Earth's gravity drags it down into a fiery reentry. Crashes do happen, but very rarely.

Is the sun just like the stars?

The sun *is* a star; what's more, it's *our* star. Earth and all the planets were formed out of the same swirling cloud of gas, 99 percent of which fell together and became the sun. The remaining 1 percent condensed into the planets, held in orbit by the new star's gravitational attraction. The sun is three hundred thousand times closer than the next nearest star. But if its distance to us were to change by just 2 or 3 percent, all life on Earth would be drowned, scorched, or frozen to death.

If the planets and the sun were all made of the same material, why aren't Earth and the other planets burning like the sun?

This is simply because none of the planets are massive enough. Nuclear reactions can take place only in conditions of extreme

Collision with a small piece of debris could do serious damage to a satellite.

pressure and temperature. None of the planets are large enough to have a gravitational pull as powerful as the sun's. Not only does the sun pull the planets toward it, it also pulls itself toward its own center.

Planets don't have sufficient mass to pull themselves in tightly enough to create the extraordinary pressures and temperatures necessary for a nuclear reaction. A nuclear reaction is similar to the way a diesel engine works. Here, a piston pushes the gas particles within a cylinder so tightly together that they get very hot. When the temperature is high enough, it explodes, forcing the piston back out.

It is the mass of an astronomical object, more than any other quality, that determines what it will become. And it is for this reason that our sun is unremarkable from the perspective of the rest of the universe, because it is actually just an average-size star. Just as a lack of mass keeps planets from becoming stars, our sun's lack of mass will determine its destiny as well.

Unlike larger stars, it will never have the gravitational pull to become a black hole, so powerful that even light is sucked into it. Instead, as the sun continues to consume itself, the outward force of nuclear explosions will overcome the gravitational power of its dwindling mass. It will expand beyond the orbit of the Earth into a red giant, huge, cooling, and dying.

Why is there life on Earth and nowhere else?

Fifty years ago, the suggestion of life on other planets raised eyebrows in both the religious and scientific communities. Now scientists believe that the universe is teeming with life, and most of the critical components seem to be found wherever we look.

But it's intelligent life that we're curious about, and this is unlikely to be found elsewhere in our solar system, simply because the stable environment necessary for life as we know it—solid landforms, an atmosphere, slight variations in temperature and pressure, light, water, and gravity—simply doesn't exist anywhere else in our solar system.

Mercury is hellishly close to the sun. Venus has days lasting eight months. Jupiter has no surface to speak of. Neptune sees the sun no more brightly than we do the moon. On Earth itself, life exists only within a very thin and precious layer. Though the layer reaches from the ocean deep to the roof of the world, this is merely 10 miles of the 3,500 that drill to the Earth's core.

The conditions for advanced life are hard to come by, and it is only by good fortune or Providence that we exist at all. On the other hand, nearly half of all stars may have solar systems, thrown out from a swirling cloud just as we were. It stands to reason that with one hundred billion stars in our galaxy and perhaps a billion galaxies, there would be no shortage of planets that meet the conditions necessary for advanced life.

The problem is that our own galaxy is a big place. With our present knowledge of physics, a round-trip to the nearest star would take thousands of years. Even with the time travel physicists believe possible, no reasonable exchange could be made with another species of life. Which is to say that if intelligent life visits us from outer space, it must be a form of life more advanced than our own, at least scientifically.

Are the stars and constellations always the same?

Yes and no. A person will never be able to see the stars move in his or her lifetime. Yet the stars are always in motion, moving thousands of miles an hour. But they're so far away that it would take hundreds of years of painstaking observations to notice, which is exactly what men have done. Why? Because people have attached great importance to the stars since the beginning of time, and records of star positions have been kept before 1000 B.C. Many constellations that are bold and bright today were mere scatterings of stars when man first plotted their positions.

On the other hand, if you travel to Australia, you will see almost an entirely different set of stars. This is simply because the Earth itself blocks our view toward the south. Which is to say that there's an entire set of stars and constellations down there, such as the Southern Cross, Argo the ship, and Corvus the crow, that those people walking upside down call their own.

It was so long before western eyes ever saw these stars that most were given names that had nothing to do with Greek mythology. These names reflected the passions of the Renaissance and Enlightenment, like Horolgium the clock, Sextant the star finder, and the Montgolfier Balloon (named for man's first flight). But these names symbolize the western conquest of the world as much as anything. Every culture on Earth has found stories in the sky, stories that endured for centuries.

In this respect, although people have always seen mostly the same groups of stars, they have interpreted them differently. Even among westerners, the number and shapes of constellations have always varied. Somebody saw a vacuum pump, another a Greek charioteer dressed in Russian winter mufti, another Noah's Ark. The Christian faithful tried to replace the twelve signs of the zodiac with the twelve apostles. Even St. Peter would have taken exception to this, as it smacked of the same paganism that replaced the celebration of the rebirth of nature with Christmas.

If the moon is spinning, why do we always see the same face? Shouldn't we be seeing the "dark side" of the moon as well?

This question puzzled men for years and was one reason men believed that the Earth, as well as the moon, was flat. If the moon was a sphere and rotated, they logically reasoned that we should see more of it.

Once again, the answer is gravity. The great mass of Earth pulls so strongly on the moon that she not only holds it close by her side, but slows its spinning as well. While the moon pulls the tides and has made Earth a slightly flattened sphere (wider at the equator), the Earth has pulled the moon into an oval shape, lopsided toward the Earth. So it is that we never see an entire side of the moon, and no one living there would have any idea of our existence.

This is called synchronous rotation, and almost every planet with moons exerts this power over them. Synchronous rotation led to the idea of a "dark side" of the moon, a poor description. The far side of the moon actually gets just as much light as the side facing Earth, and when there's only a thin crescent moon, the far side is receiving the remaining circle of light. When we cannot see the moon at all, day or night, all this means is that the moon itself blocks the light cast by the sun. The side facing us is in total darkness and is thus invisible; the far side is awash in light but turned away completely.

Why don't we see beautiful, colorful views of the sky like the pictures in magazines and on television?

Except for a few objects, *nobody* sees the universe that way, even through the world's largest telescope. This is for three reasons.

1. If you were somehow looking through the Hubble space telescope, you still couldn't see what it sees, for the simple reason that your eye has no memory. The Hubble's camera shutter can be left open indefinitely, accumulating light as a picture over time. (For the same reason, what the eye sees as motion, a camera will see as a blur.)

2. A lot of observation is done today with radio telescopes. The human eye can perceive only a small range of the available wavelengths of light (just as we cannot hear all that a dog hears nor feel the ink on a poster). A radio telescope can "see" these and record them as images as well. But no person ever actually *sees* it.

3. The photographs are colorized to show different wavelengths of light that the human eye could never see. Scientists and artists like to give them very dramatic colors.

How have people tried to use the zodiac to tell their future?

We have seen five of the twelve constellations of the zodiac: Leo, Taurus, Sagittarius, Scorpio, and Libra. These five constellations and seven others were chosen because they all lie directly in the path of the sun, moon, and planets (the plane of the ecliptic).

If you were born "under" the sign of Aquarius, it means that on your birthday the sun was in the constellation Aquarius on its journey through the sky. The sun is too bright to see the constellations behind it, so it puzzles people how anyone would know where it is. The ancients were good at estimating where it should be, because they memorized the order of the signs. They saw what sign emerged from the deepening darkness that swallowed the sun's fading light. They could then accurately estimate how many signs ahead the sun was.

They were obsessed with knowing the exact positions of the heavenly bodies, however, and built elaborate observatories to make careful records of every position and motion. This established a valuable calendar for agriculture and laid the foundations for the science of astronomy. At the same time it enabled the priesthood to make astonishing (and useless) predictions about events in the sky that were essential to maintain control of the masses. This laid the foundation for the practice of astrology, the ability to foretell the future based on the movements of the stars.

Unfortunately, they were unaware that there were three other planets, Uranus, Neptune, and Pluto, and the asteroid belt. Neither did they understand that the sun and the moon were of vastly different size, or even that all the planets went around the sun, not the Earth.

Let's assume, however, that they still had success in seeing the future from atop their magnificent observatories. It would stand to reason that, since astrologers today know all these things, they would have even better results. This might be reasonable if it weren't for the fact that, in the thousands of years that have elapsed since the birth of astrology, the Earth wobbled on its axis, like a slowing top. This resulted in everything shifting two zodiacal constellations forward. The sun's path barely goes through Scorpio anymore because of this, and spends a lot of time in nonzodiacal constellations. Present-day astrologers have not allowed for this.

Even if the ancients were really able to see the future in the sky, and even if present-day astrologers allowed for all their errors, they're still making pronouncements about daily events that are two months completely out of synchronization. It would be like getting February's weather forecast in April. It could be 100 percent accurate, but who cares?

How large is the universe?

Even if there were an answer to this question, no one would be able to truly understand it. Let's use an example of something we do know: A galaxy is generally composed of 100 billion stars. Besides the stars that we see as our own Milky Way galaxy, there's

The Andromeda galaxy. Though filled with over 100 billion stars, it is so far away as to be no brighter than a faint star. In our great universe, it is among the closer objects.

only one other galaxy we can see from the Northern Hemisphere with the naked eye, in the constellation Andromeda. It's so far away and faint that its 100 billion stars resemble only the faintest single star we can see close by in the Milky Way. This is the closest galaxy in the entire universe, the only one close enough to see. And it's estimated there are a lot more galaxies than people on Earth.

That's how large the universe is.

OBSERVATORIES AND OTHER RESOURCES

Unlike most sciences, private funding cannot begin to foot the tremendous costs of research in the heavens. The more people who appreciate the importance of stars, the more likely such research will receive government support. There are hundreds of observatories and organizations that are eager to help you learn more.

OBSERVATORIES

The Very Large Array in Socorro, New Mexico (505-835-7302), is one of many observatories that welcomes people to ask questions and walk around, free of charge. This extraordinary telescope doesn't quite match people's expectations of what an observatory should look like. Far from being a big dome with a tube sticking out of it, The Very Large Array consists of thirty satellite dishes the size of a house arranged in a huge Y pattern that expands and contracts. (They move when your not looking.) You'll have to go there to find out why, and you'll be glad you did. Like many observatories, it's in the middle of nowhere, which is to say, in beautiful country with many things to look at—like the stars. Other fascinating places are:

Lowell Observatory, Flagstaff, Arizona (520-774-3358)

Lick Observatory, Santa Cruz, California (408-459-2513)

Harvard-Smithsonian Center for Astrophysics, Cambridge, Massachussetts (617-495-7461)

Yerkes Observatory, Williams Bay, Wisconsin (414-245-5555)

William J. McCallin Planetarium, Hamilton, Ontario, Canada (905-525-9140)

Try to include one of these places on your next road trip. You will be impressed by what you see and just as impressed by astronomers' rigorous dedication to this science. If it appears that this list unduly stresses deep-space observatories over naked-eye astronomy, I would direct you to the Northwest Territories of Canada—you can find meteorites and auroras there—the Desert Southwest, your closest ocean, or any place far from light pollution, an increasingly disruptive factor to evening observations of any kind.

The Astronomical Society of Kansas City, a nonprofit group of over 250 astronomy enthusiasts, is a good example of an organization that can help direct your search for information about the stars. *Their* facility has a big dome with a tube sticking out of it. There are several hundred other such organizations around the country, so many, in fact, it would require another book to list them all. In addition to having their own observatories away from city lights, these organizations host star parties and offer educational talks on a wide range of topics. Most of these groups can be found in the Yellow Pages under "Astronomy" or "Planetarium." An observatory or organization need not be close by to have stargazing information that is useful in your location.

GEAR

A good compass can be purchased for any amount from $15 to $150. But wait, stop! You don't need a $150 compass, or even a $15 compass; just one that will point in the general direction of north. Grab something for a few dollars in a department store.

Although outdoor outfitters most often sell the more expensive compasses, they are a good source of topographic maps.

These maps are a great way to orient yourself to your surroundings. They'll show you the street you live on, the field you played ball in, and an abandoned mine shaft over yonder.

Because each star enters and departs the sky at the same place, knowing your local landmarks will familiarize you with our distant but constant companions. In popular outdoor areas, departments stores and convenience stores will sell "topo" maps. You can also order them directly from the people who make them, the United States Geological Survey, for $5 or less. Call 800-HELP-MAP or 800-USA-MAPS, or write them at USGS Information Service, Box 25286, Denver, CO 80225.

Star dials are plastic or cardboard wheels that display the constellations in season. They're a great tool to help you develop your knowledge of the sky. They can be found at planetariums, toy and hobby stores, and specialty stores like science and nature shops. A good pocket-sized star dial can be purchased for under $7. Don't bother with the glow-in-the-dark ones.

Inexpensive flashlights with red lenses can be found in sporting goods departments and military surplus stores. Look for the olive-green gooseneck type with an assortment of colored lenses.

OTHER READING

For more of the Greek and Roman sky myths, I would direct you to the timeless and readable *Bulfinch's Mythology* (Dell, 1959) or Edith Hamilton's *Mythology* (Little, Brown, & Co., 1942). I do not wish to neglect the rich and wonderful skysagas of other cultures, but their place among the stars is given only spottily in most anthologies. Though such compilations may only make reference to "our" North Star and Bears, you may be certain that every star in the heavens had a place in the oral traditions recited beneath every tree on Earth.

For more information on using the sun, moon, and stars to navigate and tell time, I would suggest *Nature Is Your Guide* by Harold Gatty (Dutton, 1958). Most other sources include helpful hints but often omit explanations of the principles you need to understand their application ("point your watch's hour hand at the sun") without which you'll either be extremely lucky in obtaining useful results, or very confused. Some sources are simply wrong. The techniques suggested in chapter 3 and the regular motions of the heavens described in chapter 1 lay the groundwork for you to understand how to tell time, navigate ,and devise techniques of your own.

For books that underscore the impact of stargazing on history, I strongly recommend

- Two truly masterful works by Lewis Mumford, *Technics and Civilization* (1967, Harcourt Brace Jovanovich) and *The Pentagon of Power* (1970, Harcourt Brace Jovanovich). Mumford explains how man's eye on the sky afforded not only advances in technology and exploration, but the means by which we have ordered our thinking and society.
- *The Haven-Finding Art* by E. G. R. Taylor (1971, American Elseier). A thorough, practical (and dry) history of man's attempts to navigate the seas, using the only constant features around him, the lights in the sky.
- *Stonehenge Decoded* by Gerald Hawkins (1965, Doubleday). Hawkins's discussion of the construction of Stonehenge as an astronomical computer is a fascinating revelation of the obsessions, intelligence, and cynicism of prehistoric man.
- *Discovering Astronomy* by Stephen Shawl (1995, Wiley). Yep, it's a textbook, which is exactly what you want. This is an exhaustive review of the basics, written by a professor more dedicated to the advancement of his students' knowledge than any others I have known.

There are two magazines about astronomy that are commonly available, with a strong popular focus. These are *Sky and Telescope* and *Astronomy*. Each year in December they each publish an issue that lists all the clubs, organizations, and observatories in the United States.

The Farmer's Almanac is another good source of information on such things as the location of planets, phases of the moon, and comet pathways, as are the larger almanacs like *World* and *New York Times*.

Many newspapers give information on positions of the planets, phases of the moon, tidal activity, and sun and moon risings and settings in the weather section.

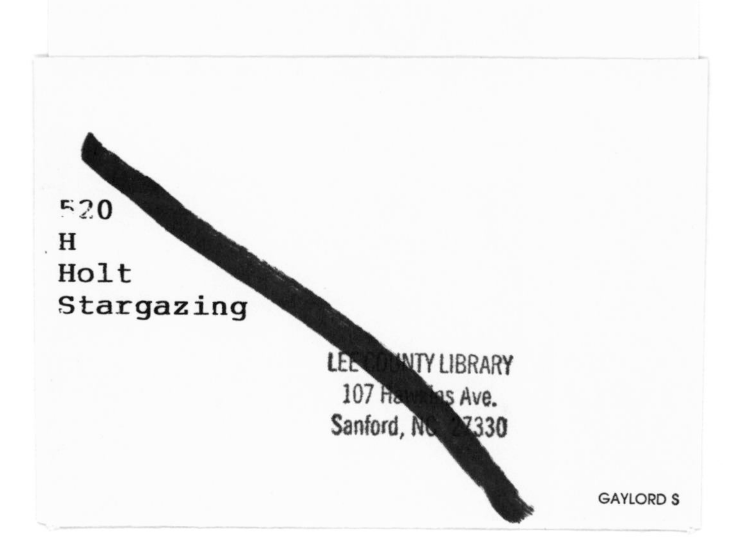